建筑立场系列丛书 No.56
C3
U0351520
从教育角度看幼儿园建筑
Buildings for Kids as Educators
汉英对照
(韩语版第372期)
韩国C3出版公社 | 编
史虹涛 时真姝 马莉 栾一斐 张琳娜 陈帅甫 蒋丽 | 译
大连理工大学出版社

C3 建筑立场系列丛书 No. 56

4 小型度假屋

54 从教育角度看幼儿园建筑

126 城市住宅 后社区时代

C3 No. 56 Buildings for Kids as Educators

Longbush生态庇护所迎客小屋_Sarosh Mulla Design

Longbush生态庇护所迎客小屋是一个创新型环境教育空间，其设计、建造和运营都由一群热情洋溢的志愿者在当地企业和慈善机构的支持下完成。所有游客均可免费来访迎客小屋，该项目旨在促进我们对保护区和城市自然环境的积极管理。

该项目是由获奖设计师Sarosh Mulla发起并亲自设计的。Mulla带领一支由88名志愿者组成的队伍完成了空间的建造工作，为来访的学校团体、生态学家和游客提供设施。小屋的结构设计借鉴了风景画派的框架技术，但是采用当代的表现形式，以促进小屋与环境的积极融合。不仅是简单地提供观景之地而已，迎客小屋还通过各种各样的项目，鼓励游客参与生态庇护所的环境复原工作。

建筑采用宽大的钢结构和织物搭成遮棚，为建在山坡上的户外教室遮阳挡雨。遮棚下方的三个木质围合结构和几个小花园共同构建了这处户外教室。这一设计的内涵在于使用自然材料与合成材料，共同作用于生态修复的环境之中。

每个木质围合结构都分别提供了不同的服务，包括教学资料的存储、洗浴设施以及当地生态学家的办公室。办公室通过一条巨大的吊桥向外界敞开，吊桥放下之后可以形成一个水平的平台，用于教学和演示活动。另一个木质围合结构设有平面屋顶，站在屋顶甲板上，山谷美景一览无余。人们可以爬上木梯到达这个屋顶，梯子采用就地采伐的麦卢卡树制成。

生态庇护所的面积大约有120ha，经过15年的不懈努力，生态环境已经得到迅速恢复。通过清除有害动植物和杂草、种植几十万株本地树木等措施，Longbush地区多样化的生态环境焕发出了新的生机。在新西兰的历史上，从未有过由如此庞大而多样化的志愿者和赞助商群体共同建造的公共建筑。迎客小屋对当地的环境需求做出了积极的响应，为高品质的社群共建建筑打造了一个新的标杆。

Longbush Ecosanctuary Welcome Shelter

The Longbush Ecosanctuary Welcome Shelter is an innovative environmental education space designed, constructed and operated by a group of passionate volunteers, with the support of local businesses and charitable organizations. Access to the Welcome Shelter is free of charge for all visitors and the project aims to promote active stewardship of our natural environment both in conservation areas and in our cities.

The project was generated and designed by the award winning designer Sarosh Mulla. Mulla has led a team of 88 volunteers in the construction of the space that provides facilities for visiting school groups, ecologists and tourists. The design of the structure draws on the framing techniques of the Picturesque, but applies these through contemporary forms, which promote active engagement with the environment. Rather than simply viewing the landscape, visitors are encouraged to take part in the environmental restoration occurring at the ecosanctuary through the programs offered

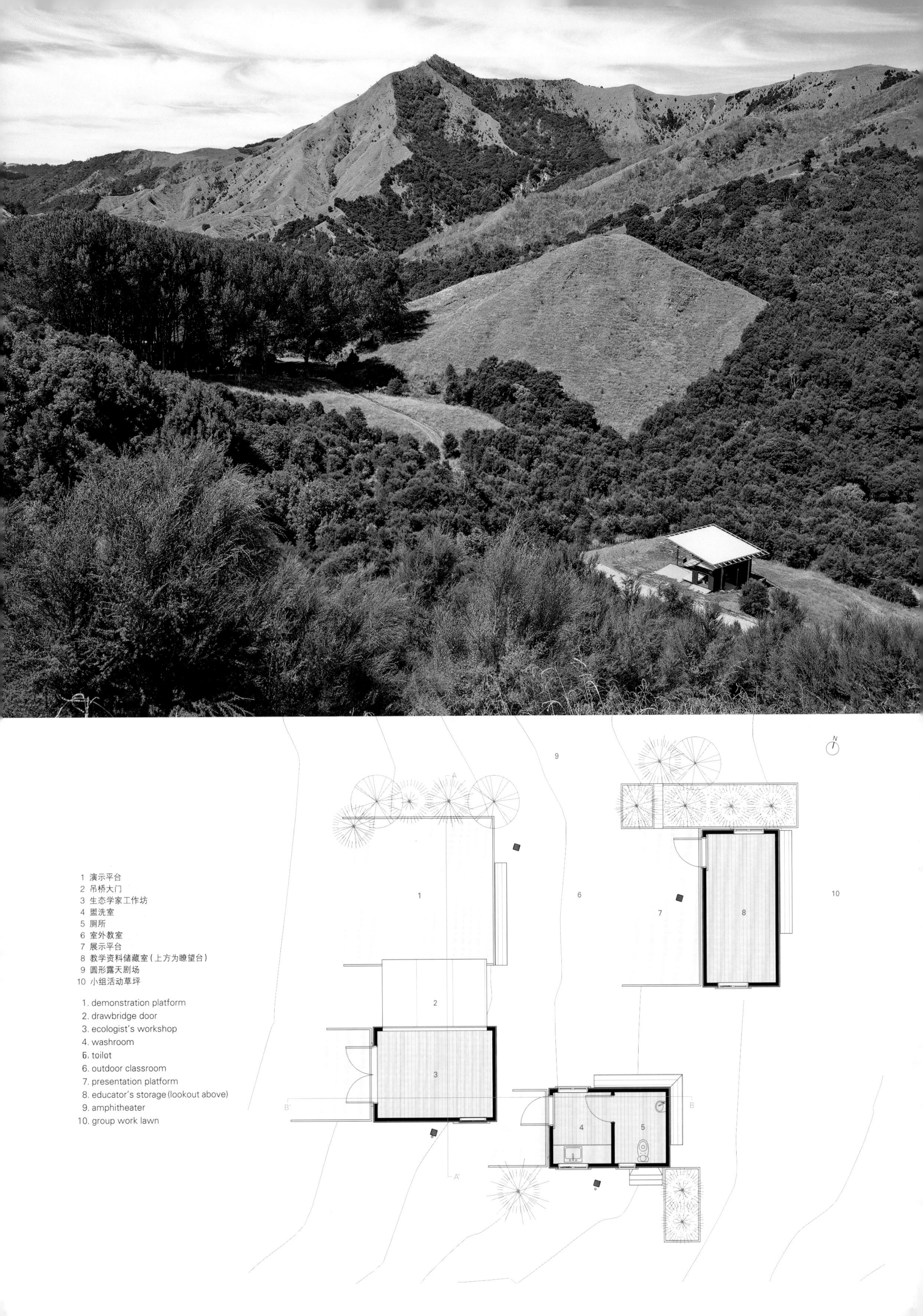
1 演示平台
2 吊桥大门
3 生态学家工作坊
4 盥洗室
5 厕所
6 室外教室
7 展示平台
8 教学资料储藏室(上方为瞭望台)
9 圆形露天剧场
10 小组活动草坪
1. demonstration platform
2. drawbridge door
3. ecologist's workshop
4. washroom
5. toilot
6. outdoor classroom
7. presentation platform
8. educator's storage (lookout above)
9. amphitheater
10. group work lawn

at the Welcome Shelter.

A large steel and fabric canopy provides shelter from the sun and rain for an outdoor classroom created on the hillside. Below the roof the form of this classroom is defined by the position of three timber enclosures and small retained gardens. The design plays with the connotations of natural and synthetic materials in the setting of the recovering environment.

Each timber enclosure provides a different service, including the storage of teaching material, ablution facilities and an office for the site ecologist. This office opens up to the exterior with a large drawbridge to create a level platform on teaching and demonstration days. Another enclosure includes a roof deck from which a stunning view of the valley can be appreciated. This deck is accessed via a traditionally made greenwood ladder made of a Manuka tree harvested from the building site.

The ecosanctuary is approximately 120 hectares and over the past 15 years has been the rapidly restored through the efforts. With the removal of invasive pests and weeds, alongside the planting of hundreds of thousands of native trees, the diverse ecology at Longbush is beginning to thrive again. Never before in New Zealand history has such a large and diverse group of volunteers and sponsors created a piece of public architecture. The Welcome Shelter creates a new benchmark for high quality community generated architecture that responds to the needs of the local environment.

项目名称：Longbush Ecosanctuary Welcome Shelter
地点：Gisborne, New Zealand
建筑师：Sarosh Mulla Design
项目团队：88 Community volunteers, 88 sponsors
客户：Longbush Ecological Trust
建筑面积：150m²
尺寸：12m x 12m x 6m
施工时间：2013.10—2014.12
摄影师：©Simon Devitt (courtesy of the architect)

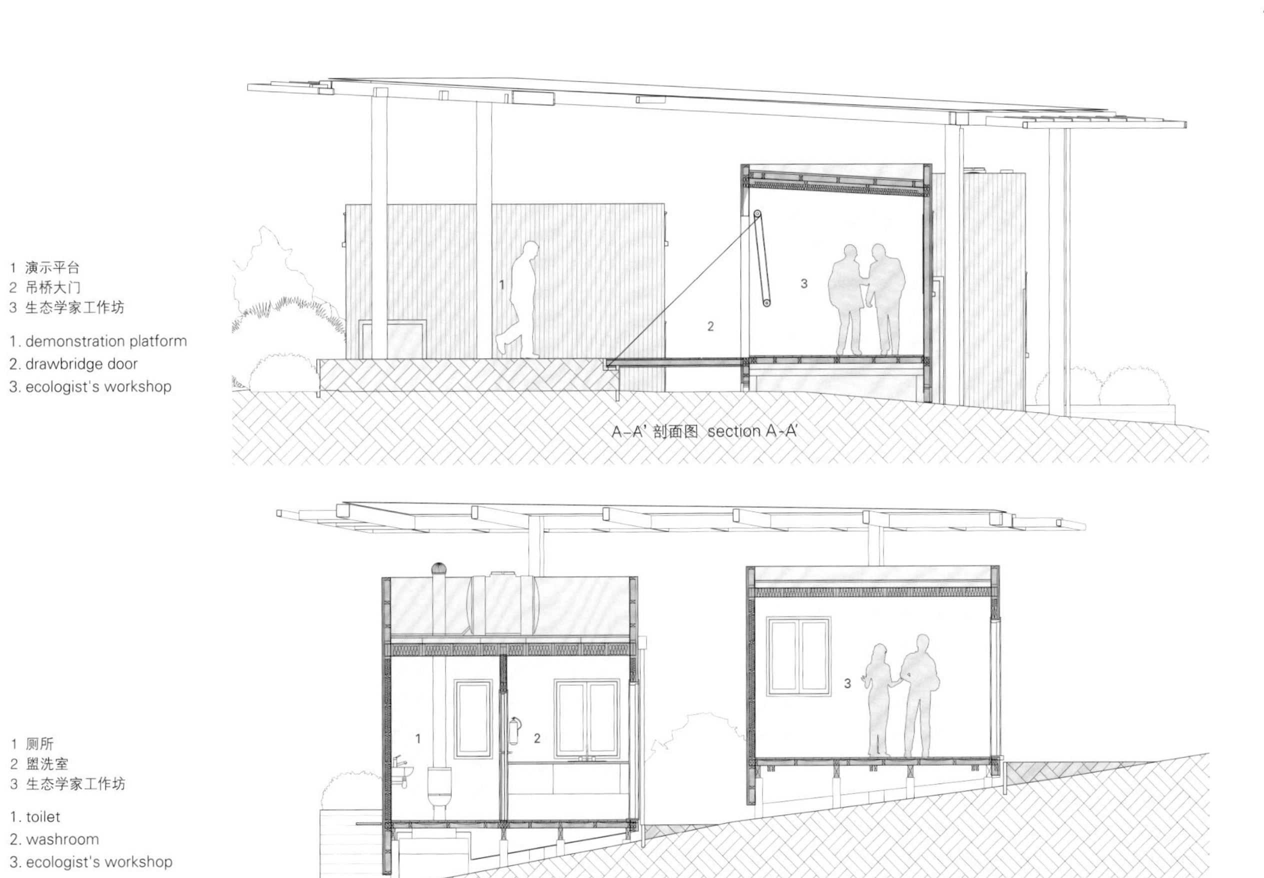

1 演示平台
2 吊桥大门
3 生态学家工作坊
1. demonstration platform
2. drawbridge door
3. ecologist's workshop
A-A' 剖面图 section A-A'
1 厕所
2 盥洗室
3 生态学家工作坊
1. toilet
2. washroom
3. ecologist's workshop
B-B' 剖面图 section B-B'

Tubakuba山脉小墅 _OPA Form

在卑尔根最著名的山脉的森林中，经过一条喇叭形状的隧道，就能进入一个高悬在城市之上、隐秘在树木之后的"木质泡泡"中。在这里，你可以度过一晚只属于你和孩子们的欢乐时光。该项目由OPA建筑事务所的建筑师Espen Folgerø主持，联合卑尔根建筑学院的设计建造工作室共同打造完成。

Tubakuba项目有95%是由木材建造而成的。内墙覆盖胶合板，并采用灵活的挪威西部松木板，外部使用落叶松木烤板。独具特色的隧道由削弯的松木堆叠而成，提供足够的强度；同时南墙上铺设原始落叶松木，其材质会随时间推移而逐渐变成灰色。烤色镀层采用日本传统的烧杉板工艺加工，这种处理方式能够预防木质霉变和腐烂。

本案没有电力支撑，是卑尔根仅有的离网型旅馆房间。项目通过缩小户内空间使热量消耗降至最低。无论是墙壁、地板还是天花板，用于构造与保温的材料都是木材。入口处弯曲的木条是锯木厂成品的副产品，镀层由碳化的（烤制的）二级镀木板制成。墙壁、屋顶和地板内的保温材料是木质纤维——一种吸湿性材料，可以让建筑透气，而不用机械通风。

在城市和野外的边界之处，Tubakuba提供了一间免费的旅馆房间。尤其针对带孩子的家庭，该项目的目的是让每个人都有机会享受户外体验，并给孩子一个在森林中过夜的美好初体验。

Tubakuba是一个带景观的14m²的房间，类型介于帐篷和木屋之间，可以通过"喇叭隧道"进入其中。你可以深切感受到漂浮在城市上空的感觉，特别是当你靠近巨大的落地窗，俯瞰窗外陡峭的山坡以及坡下的市中心的时候。在别墅下的阴凉处，能找到一个不错的野餐位置——喇叭隧道也可以作为避雨、玩乐的好地方。

除了使用灵活之外，Tubakuba还为现代城市添置了一些我们现在可能缺乏的东西：一座都市风格的小木屋。想要到达这里，你既不需要汽车也不用出行到很远的地方，因为它离公共缆车很近，缆车每隔15到30分钟便会从市中心出发，上下Fløyen山一趟。

Tubakuba Mountain Hub

Hovering above the city, hidden behind trees in the forest of Bergen's most famous mountain; through a tuba-tunnel, you can enter a wooden bubble. One night, just for you and your kids. The project is the result of a design-build workshop at Bergen School of Architecture lead by Espen Folgerø at OPA Form.

Tubakuba is constructed of 95 percent wood. The interior is clad in plywood, while flexible wooden boards of the western Norwegian pine and the exterior is clad with burned larch. The characteristic tunnel consists of curved shavings of pine mounted in layers to provide sufficient strength, while the south wall is clad with untreated larch, which will turn gray with time. The burned cladding is made with the traditional Japanese method Shou Sugi Ban. This is a treatment to prevent fungal decay and damage.

With no electricity - the project is the only off-the grid hotel-room in Bergen. The project aims to minimize the need for heating by minimizing the indoor volume. The materials

©Marina Magreøy (courtesy of the architect)

项目名称：Tubakuba Mountain Hub
地点：Bergen, Norway
建筑师：OPA Form
项目团队：professor_Espen Folgerø, Håvard Austvoll, Sigurdur Gunnarsson, Hans Christian Elstad / student_Gunnar Sørås, Bent Brørs, Ida Helen Skogstad, Adrian Højfeldt, Eivind Lechbrandt, Alice Guan, Luise Storch, Eline Moe Eidvin, Shepol Barzan, Øyvind Kristiansen, Stein Atle Juvik, Eva Bull, Kristian Bøysen, Sondre Bakken
竣工时间：2014
摄影师：©Helge Skodvin (courtesy of the architect) - p.12, p.13
©Gunnar Sørås (courtesy of the architect) - p.10, p.11 bottom

©Stine Elise Kristoffersen (courtesy of the architect)

chosen for the construction and insulation is wood all the way through the walls, floor and ceiling. The bent strips of wood in the entrance can be found as shavings from sawmill production and the cladding is made by carbonizing(burning) second grade wooden cladding planks. Inside the walls, roof and floor the insulation consists of wooden fibers–a hygroscopic material that allows for the construction to breath, excluding the need for mechanical ventilation.

The Tubakuba offers a hotel room that is free of charge on the border between the wild nature and the city. Especially aimed at families with young children, the aim of the project is to make outdoor experiences possible for everybody and give children a positive first meeting with spending the night in a forest.

Tubakuba is a 14 square meter room with a view, somewhere between a tent and a cabin, that you enter through a "tuba tunnel". The feeling of floating over the city is highly present, especially when you get closer to the large windows facing the steep hill down towards the city center. Sheltered under the cottage you will find a nice picnic area - and the tuba tunnel can function as a shelter for rain, for fun and play.

In addition to being flexible in its use, the Tubakuba adds something to the city that we may lack today: an urban cottage. To use this cottage you neither need a car nor travel very far, because of the proximity to the public tram which climbs up and down Fløyen from the city center every 15-30 minutes.

N
0
0.5
1m

湖滨小筑_FAM Architekti + Feliden+Mawson

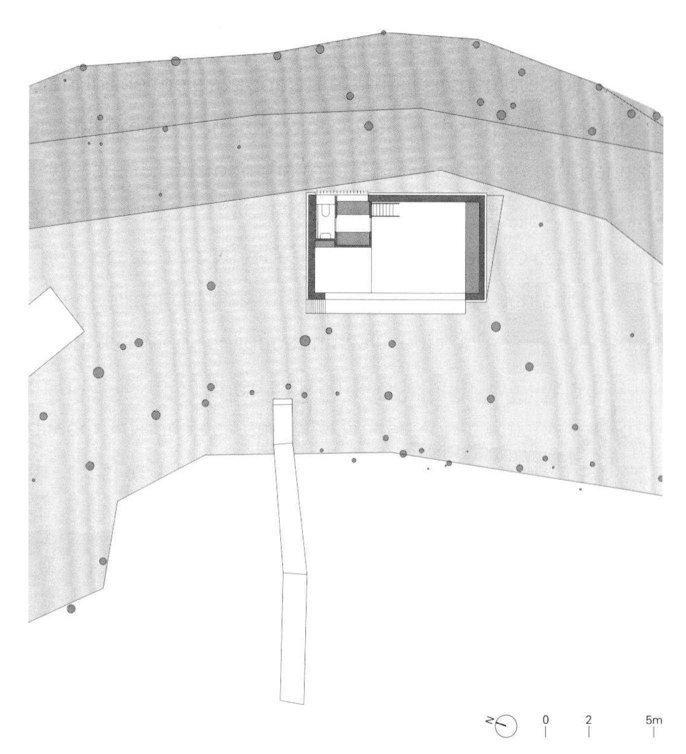

湖滨小筑位于波西米亚北部，该项目在尊重场地独特的自然环境和原址的原始结构基础上进行了重建。项目旨在为满足客户的扬帆出游兴趣提供一个全年的休憩场所，尽量减少对地形的改造，同时最大限度地建立与湖泊和周边松树林之间的视觉联系。单坡屋顶的设计既是室内空间规划的结果，也考虑了与湖滨轴心的关系问题，泊船码头代表轴心，是通向湖滨小筑的主要入口。

湖滨小筑的室内入口是一个巨幅的玻璃洞口，内部设计成与屋顶相连的连续性空间。顶部是一处就寝的睡床，正下方是一个黑盒子型的紧凑空间，内部设有极简的厨房、卫生间以及淋浴设施。室内的主墙上建有一个与小筑等长的纵深的柜橱，用作储藏空间，附带一个嵌入式壁炉。内部空间的连续性通过白色漆面的木板包层而得以加强。地板为彩色的水泥砂浆材质，与周边的沙质地面相互协调。

空间设计的重点在于既要能够看到广阔的湖景，同时还要使湖滨小筑融入到周围的自然环境之中。项目的外观采用落叶松材质的木板层，映衬出松树的挺拔。在项目的设计过程中遇到的一个实际问题是，如何确保在无人使用时小筑本身的安全问题。为此，采用完全相同的落叶松木板，设计建造成固定、可折叠的百叶窗，以保护建筑的各个入口，并在整栋建筑关闭时打造出一种延展连续的立面效果。

Lake Cabin

The replacement of an old cabin on a lake shore in Northern Bohemia respects the unique natural character of the site and follows the cabin's original outline. The brief of the project was to provide an all year round retreat for the clients' sailing passion, with minimum typology and maximum visual connections to the lake and the surrounding pine forest. The single pitch is a result of internal spatial planning as well as the relationship to the lake–shore axis represented by the mooring pier as a principle access point.

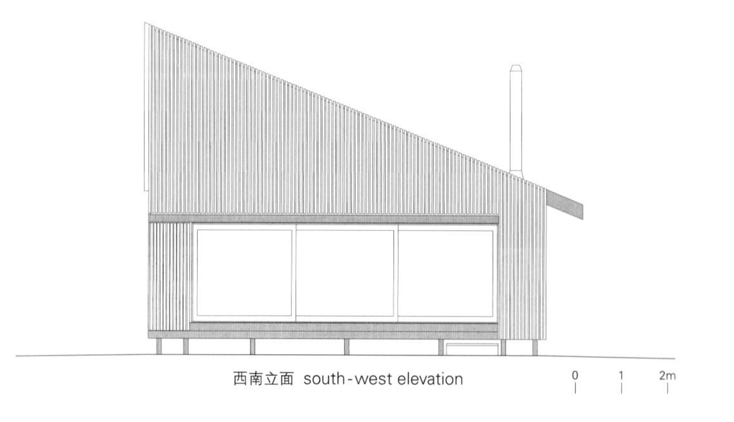

西南立面 south-west elevation

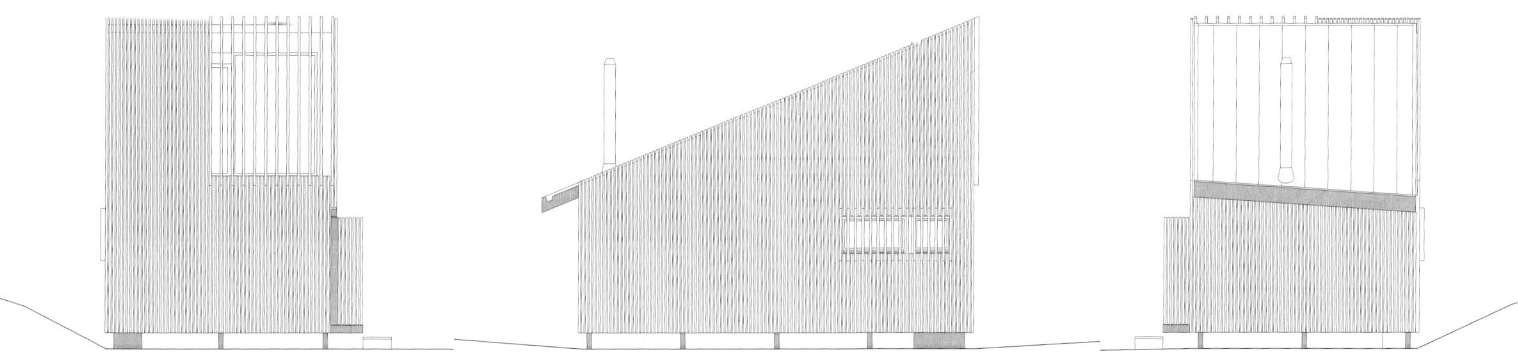

西北立面 north-west elevation

东北立面 north-east elevation

东南立面 south-east elevation

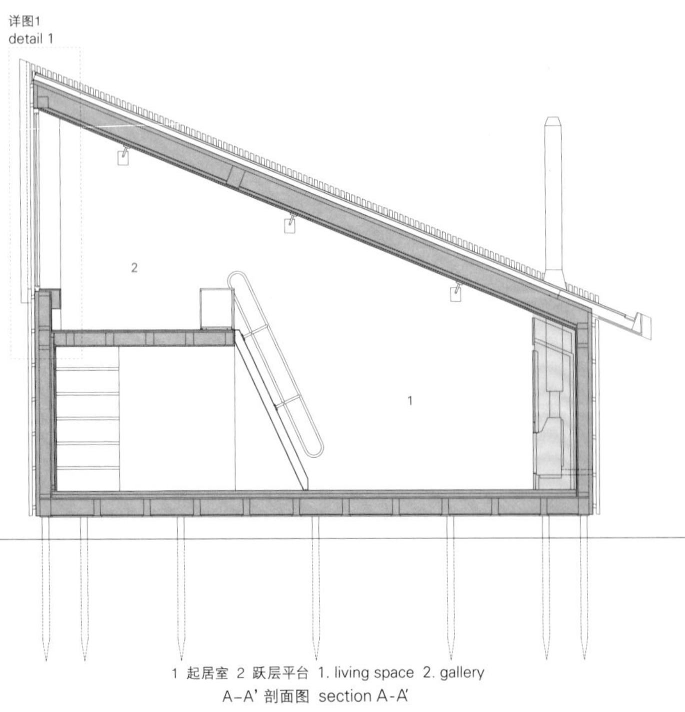

1 起居室 2 跃层平台 1. living space 2. gallery
A-A' 剖面图 section A-A'

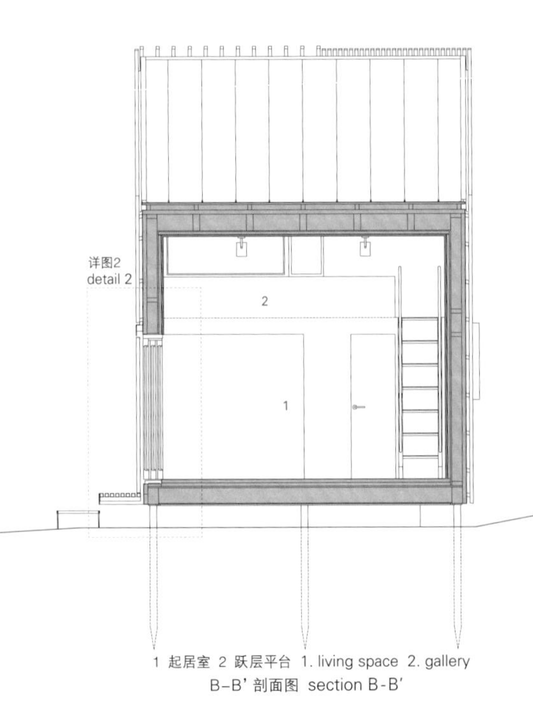

1 起居室 2 跃层平台 1. living space 2. gallery
B-B' 剖面图 section B-B'

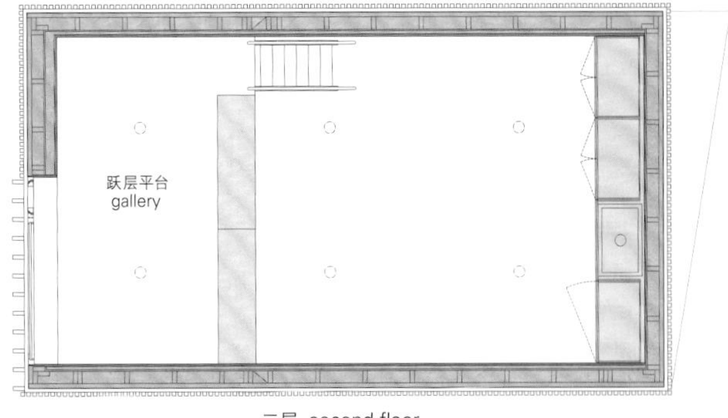

二层 second floor

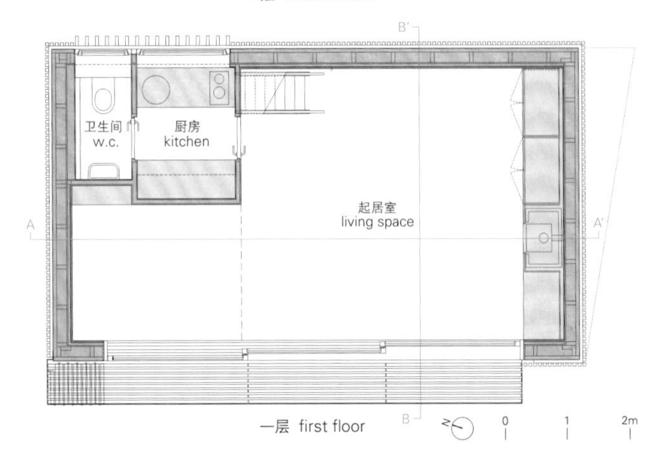

一层 first floor

The interior of the cabin is accessible through the large window opening and is designed as a continuous space open to the roof. The tall end includes a sleeping gallery with a compact black box underneath containing a minimal kitchen, a toilet and shower facilities. The principle wall of the interior forms a deep full length cupboard for storage with a built-in fire place. The continuity of the interior is enhanced by the unifying jangled timber cladding sealed in white oil finish. The floor is a sand-colored cement screed relating to the sandy ground of the beach around the cabin.

The main focus of the space is on the vast views of the lake and on the relationship of the cabin with its natural surroundings. The exterior larch cladding reflects the verticality of the pine trees. One of the practical issues of the project was the safety of the building when not in use. Fixed and folding shutters with the identical larch cladding protect the openings and when closed it provides a continuous elevation effect.

项目名称：Lake Cabin / 地点：Machovo jezero, Doksy, Czech Republic / 建筑师：FAM Arkitekti, Feilden+Mawson / 项目团队：project architect_Pavel Nasadil, Jan Horky, Ondrej Freudl / 机电工程顾问：Milan Mastalka, Zdenek Slavik, Vratislav Bilek / 工料测量师：Propos / 总承包商：Vittore, s.r.o. / 分承包商：Bernt Interiery, Zamecnictvi Tocimon / 照明顾问：Profilighting, s.r.o. / 用地面积：42.5m² / 有效楼层面积：42.9m² / 体积：199.8m³ / 热量需求：2.7kW / 建筑批准时间：2011 / 施工时间：2013.8—2014.8 / 摄影师：©Tomas Balej (courtesy of the architect)

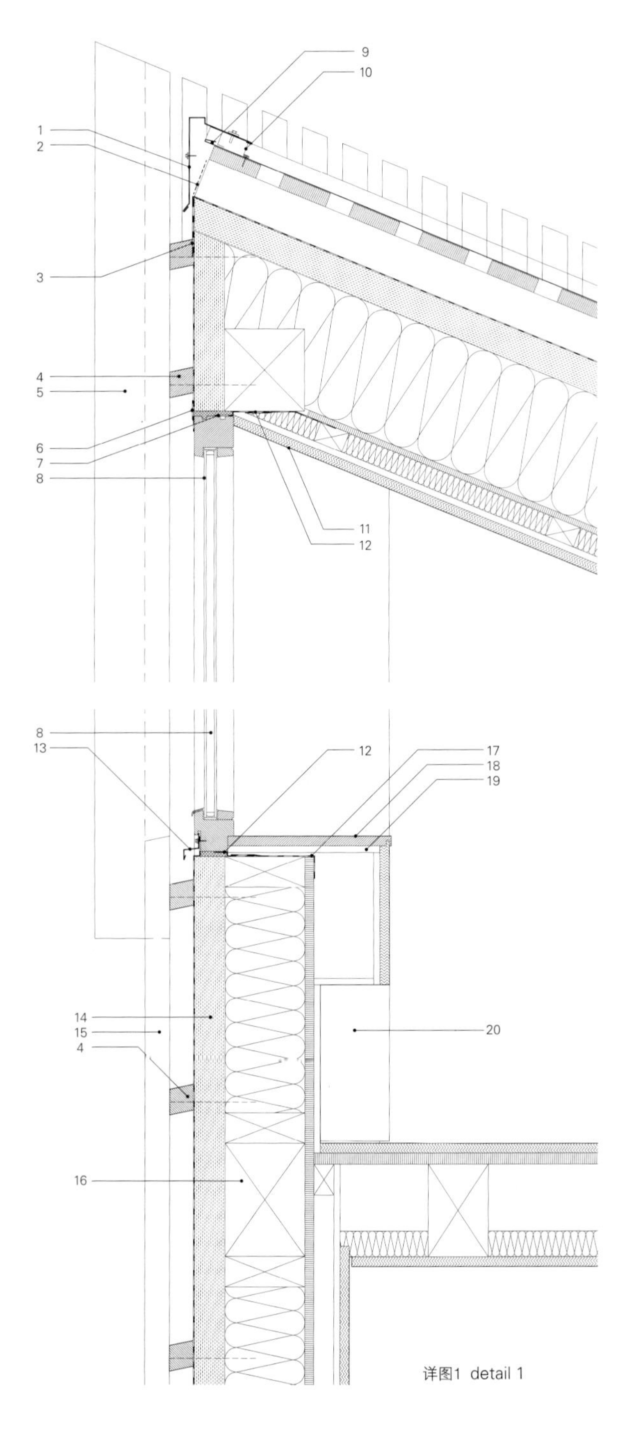

详图1 detail 1

1. covering steel plate
2. black ventilation band
3. roof sheeting
4. skewed larch counter batten 50/50
5. larch batten
6. adhesive tape for facade sheeting
7. sealed assembly joint
8. window flush with wood fibre board
9. return fold
10. vent
11. internal deck cladding
12. internal sealing tape
13. water bar
14. wood fibre board 60mm
15. timber tie frame
16. window sill to match interior decking
17. plaster board
18. niche for electric radiator
19. Purenit joist 60/150/150
20. steel bracket 90/40/3
21. steel cover sheet 2mm
22. fittings for sliding/folding windows
23. compressed band illbruck illmod
24. sliding folding shutter
25. sliding window-HS portal
26. lintel BSH 160/460
27. aluminium profile 25/25/2
28. timber joist
29. Geze guard rail
30. 2 x larch batten 60/40
31. terrace steel console
32. steel column
33. air gap
34. ledge to support HDF board
35. expansion joint
36. sealing bitumen band
37. steel plate

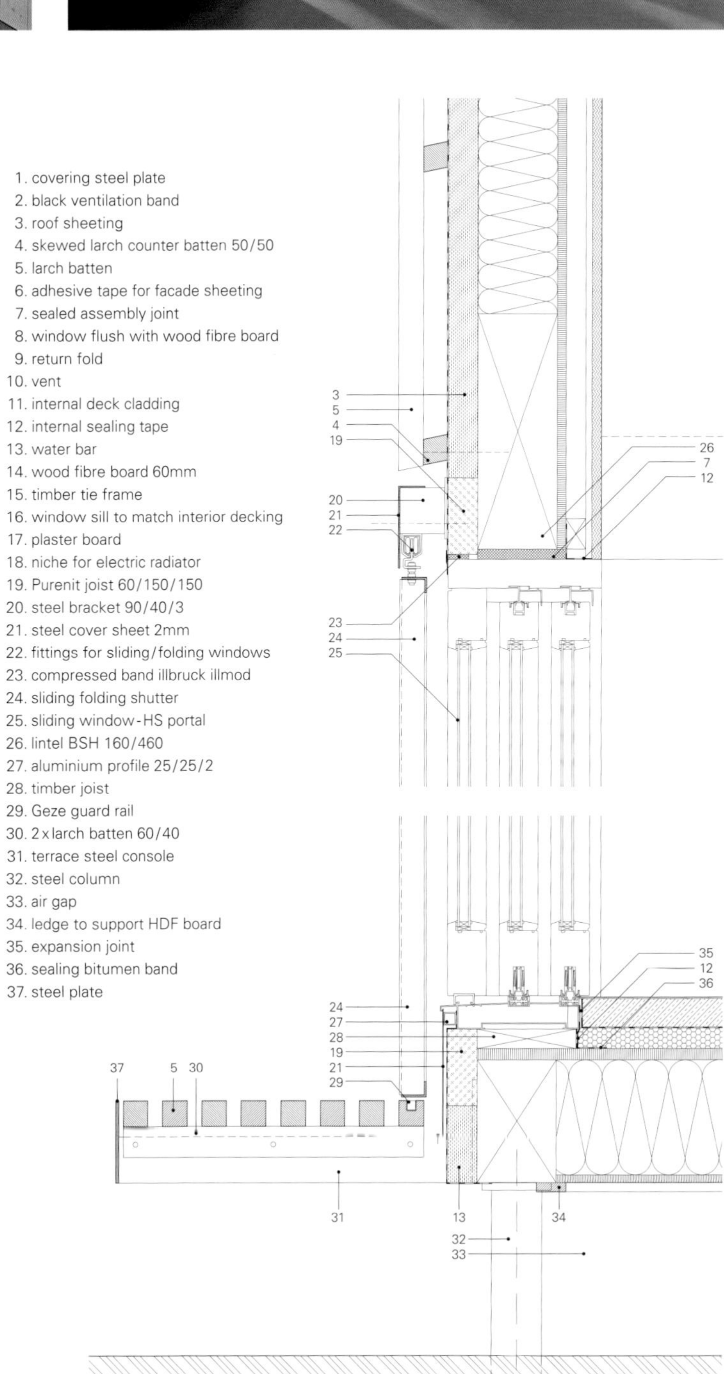

详图2 detail 2

林中小屋_Uhlik Architekti

有客户联系我们，希望能在乡下建造一处隐蔽的居所，让他可以"隐居"起来，为他在布拉格高强度的工作养精蓄锐。我们从波西米亚中部与南部之间的地区着手考虑，因为那里是这位客户的家乡，也是他从小就非常喜欢去的地方。他带我们来到其中的一处场地，这里有绿树、草地，怪石嶙峋，是一个远离尘嚣的世外桃源。奇异的景观与客户的目标需求立刻打动了我们的心。片刻的犹豫过后，我们决定自己动手，打造一栋"林中小屋"（这是我们后来对它的称呼）。建筑材料已经确定：基础材料采用来自周边森林的木材，除此之外，我们还充分利用了当地的资源和能工巧匠。

我们把整栋建筑设计成一个紧凑而封闭的体块——随意地搁置在石块之上，其中一端像船尾一样被抬高，支撑在一块巨石之上。这栋封闭的黑色建筑由碳化木材制成，内部是一个整体连通的空间，建筑规格为3.1m×5.8m。入口处是一个平台，挑高仅为一人高，这里安装了一块大幅的玻璃，可以透过玻璃观赏林中的景致。建筑的其他部分逐渐升高，通向另外一个入口，与树冠平齐。室内的休息之处非常独特，每一层台阶上都可以躺下休息。台阶下方是储物空间，将台阶的折叠处翻转过来，就成为一张双人床。室内空间是多功能的，既适合居住，也适合举办小型活动、表演，或者用来冥想也未尝不可。两个入口处的嵌入玻璃可以用大幅的百叶窗来遮挡，较大块的玻璃还配有滑轮和手摇绞车。

林中小屋的基础支撑框架是一个龙骨结构，由当地的木匠共同合作打造而成。建筑的外部覆盖着一层碳化木板，上面开有相互接合的槽口。木板和龙骨所用的木材来自于屋主人自家土地上的枯树。建筑内部铺设定向刨花板，既加固了建筑，又降低了成本。屋面覆盖了两层沥青毡。屋面的排水系统采用一条倾斜的L型钢，安全可靠，同时也充当滑轮的轨道。所有钢结构构件都参照附近村庄的一位铁匠所提供的资料来准备。整栋建筑仅用了2012年秋季至2013年春季几个周末的时间便组装完成。

Forest Retreat

Our client contacted us with an idea to create a hideaway in the countryside where he could "hole up" and gather strength

for his demanding work in Prague. We set off into an area between Central and South Bohemia where he comes from and where he has loved going to since his childhood. He took us to a spot in the midst of fields, woods and meadows, full of strange boulders, to a remote and somewhat forgotten place. The magic landscape together with the client's aim won our hearts immediately. After a short hesitation, we decided to build the "forest retreat"(as we called it later) together by our own hands. The building material was given: the base was wood from a nearby forest, for the rest, we utilized local sources and skills of local craftsmen.

We designed a compact enclosed volume – an object resting freely on boulders with a stern raised on a huge boulder. The enclosed black object made of charred wood contains one

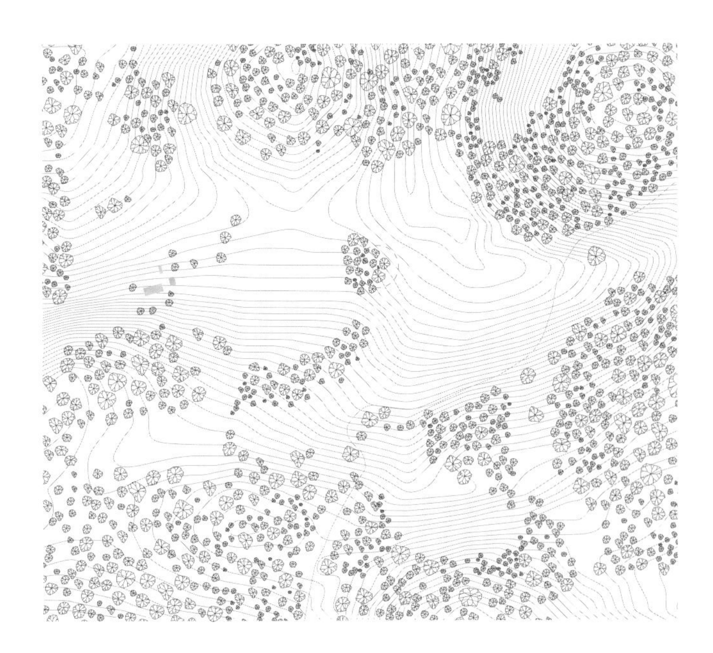

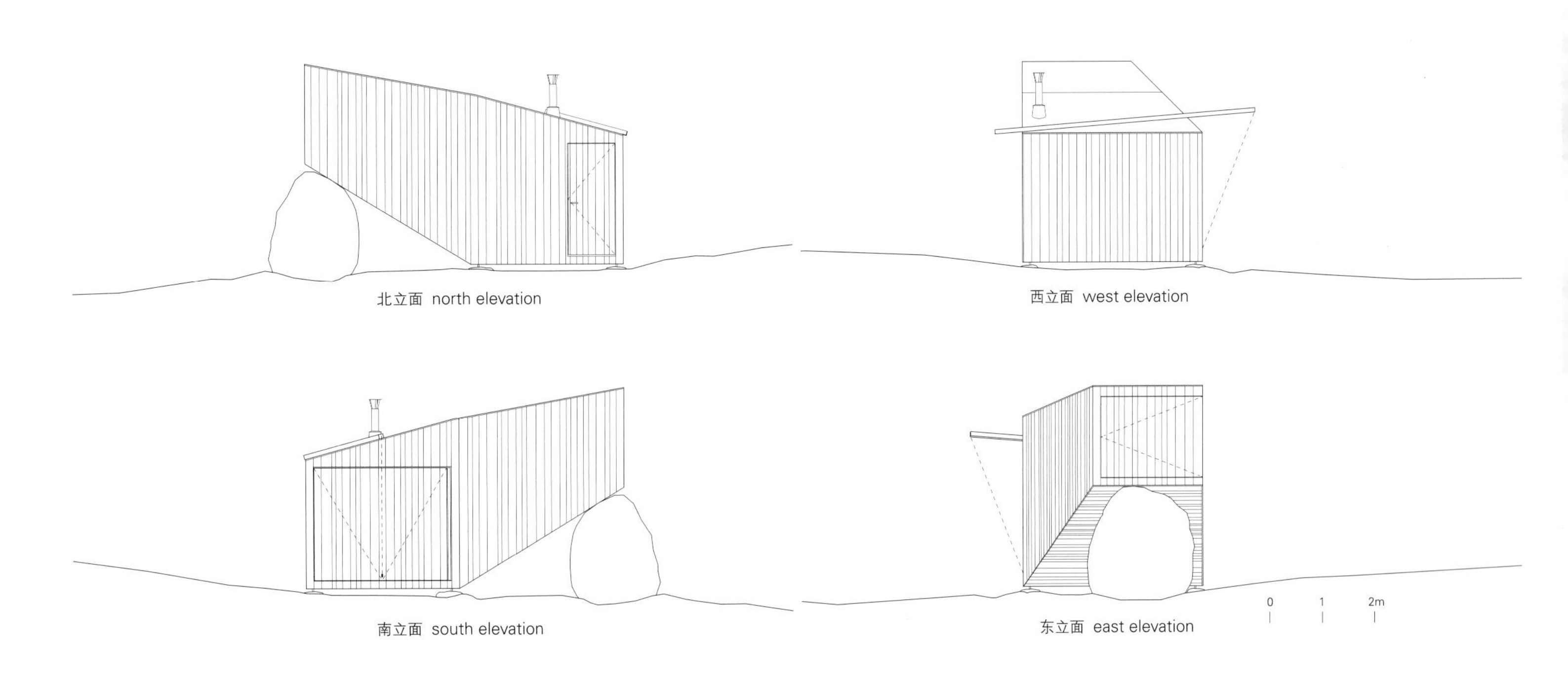

interconnected space with the dimensions of 3.1m x 5.8m. In the entering part is a flat platform with headroom just for a standing person; this part is opened by a large glassed-in surface offering the view. The remaining part of the object rises gradually towards another opening, facing the tree crowns. And there is a unique place for resting. Each step can be used as sleeping place. Under them, there is storage space; turning one of the benches over creates another double bed. The space is multifunctional, suitable not only for dwelling but also for smaller events, performances or just meditation. Both glassed-in openings can be closed by big shutters, the bigger one by a pulley and a hand winch.

The basic supporting framework of the forest retreat is a joist construction which emerged from the cooperation with local carpenters. The outside is covered with charred boards with rabbet joints. The wood for boards and joists was taken from fallen trees from the owner's land. The inside of the construction is coated with OSB boards in order to reinforce it and also with respect to the low price. The roof is covered by two layers of asphalt felts. The drainage of roof is secured by an L-profile steel, placed obliquely, which serves at the same time as a rail for attaching the idler pulley wheel. All steel components were prepared according to our documentation by a blacksmith from a nearby village. The building was assembled during a few extended weekends between the autumn 2012 and the spring 2013. Uhlik Architekti

项目名称：Forest Retreat
地点：Central Bohemian Region, Czech Republic
建筑师：Uhlik Architekti
项目团队：Petr Uhlik, Jan Sorm, Premysl Jurak
合作方：Martin Pazdernik, S.K.
工艺：Sindelar, Jaroslav Hula
用地面积：420,000m²
建筑面积：9.8m²
总建筑面积：16m²
竣工时间：2013
摄影师：©Jan Kudej (courtesy of the architect)

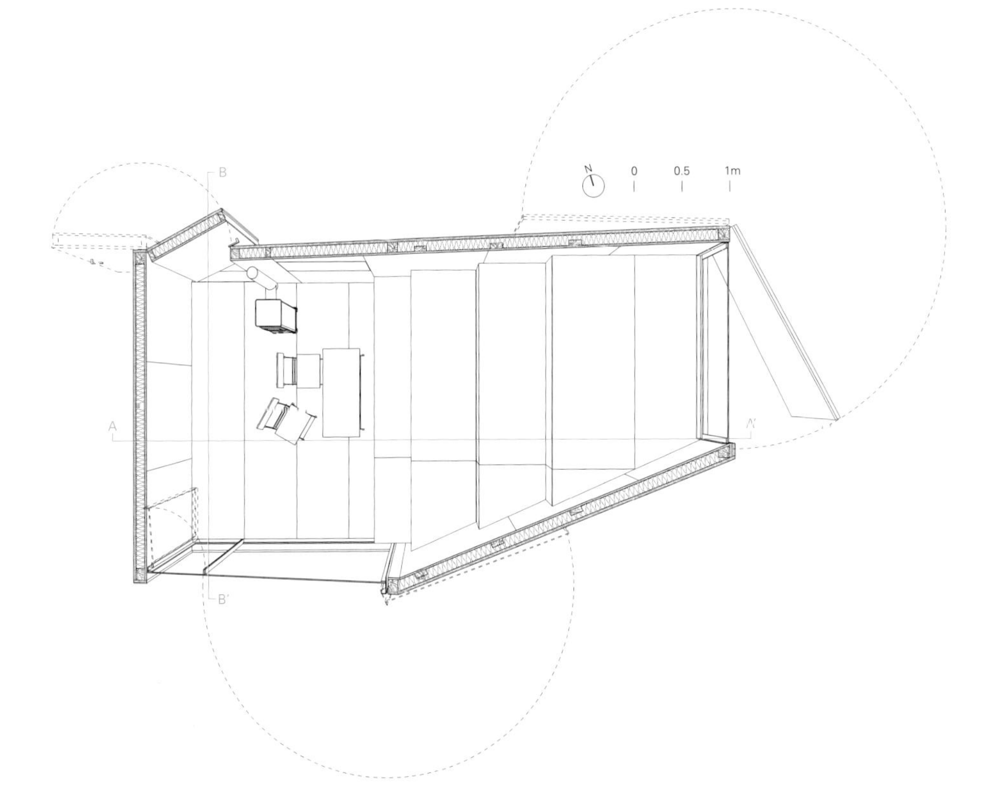

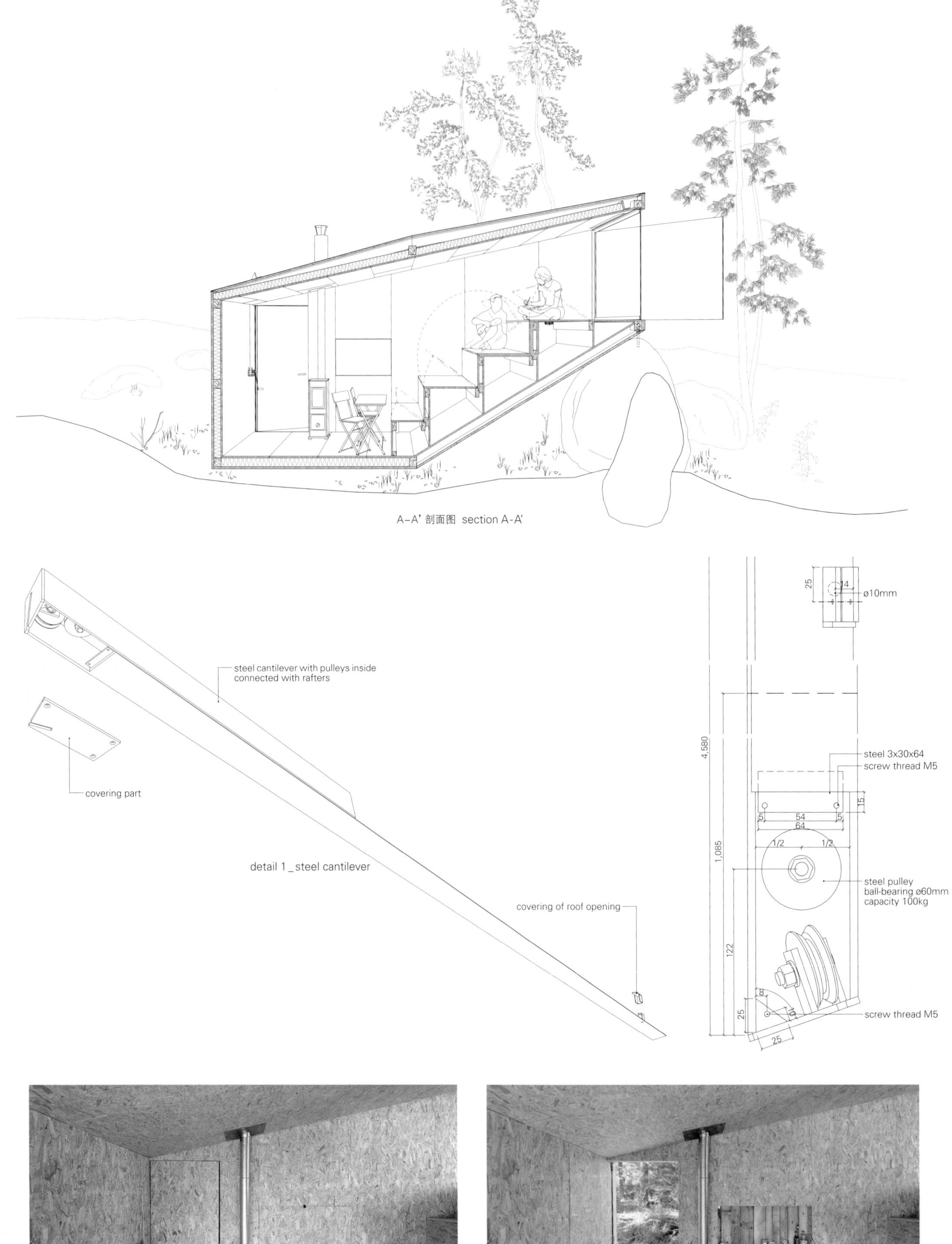
A–A' 剖面图 section A-A'
steel cantilever with pulleys inside
connected with rafters
covering part
detail 1 _ steel cantilever
covering of roof opening
25
14
ø10mm
4,580
steel 3x30x64
screw thread M5
15
5
54
64
5
1/2
1/2
1,085
steel pulley
ball-bearing ø60mm
capacity 100kg
122
8
10
25
screw thread M5
25

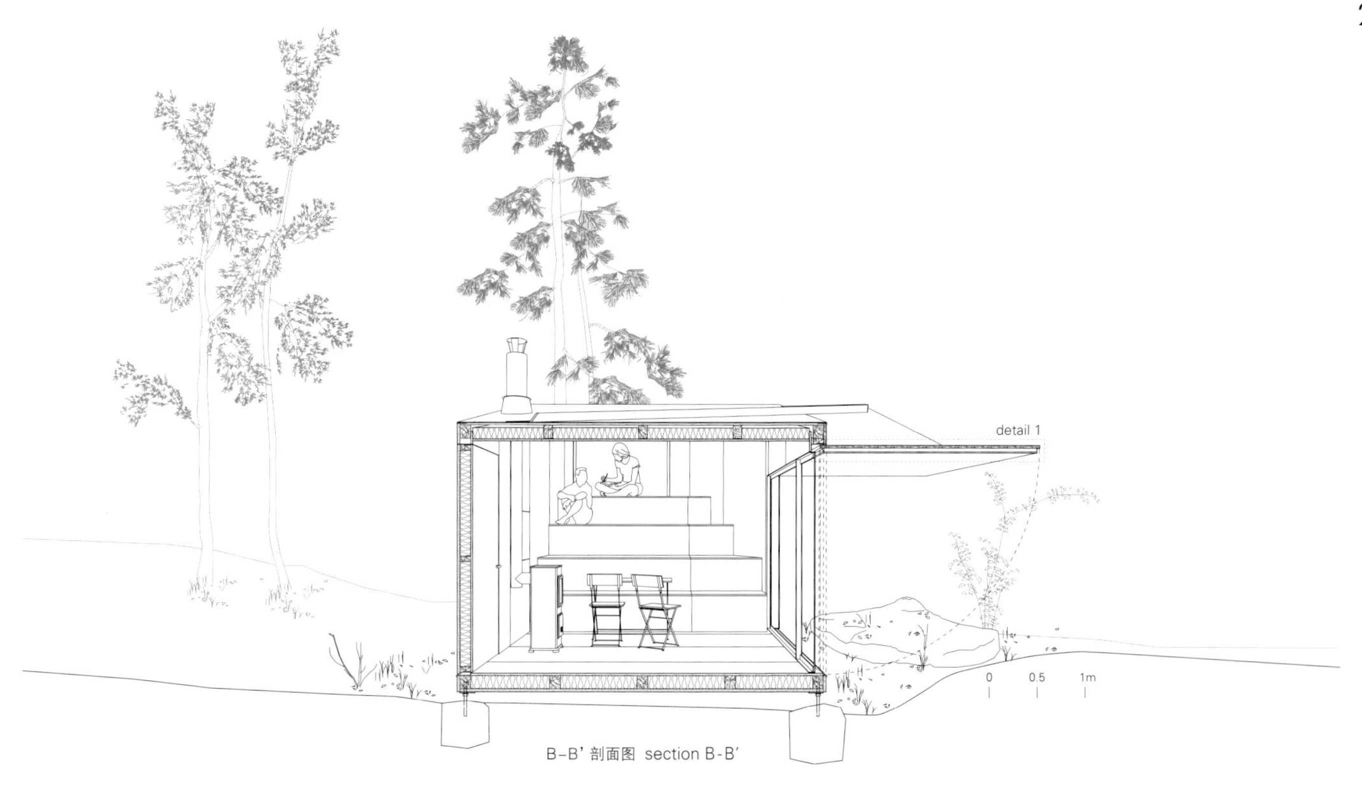

B-B' 剖面图 section B-B'

隐形住宅_Delugan Meissl Associated Architects

隐形住宅是一个灵活的住宅单元，包括一个预制组装的木结构，适合在任何指定的场地安装启用。极高的灵活性和空间品质是隐形住宅开发理念的核心要素。开放式的格局由一个玻璃罩、一块湿电池构成，构建出三个宽敞的住宅单元，以供个性化的设计和使用。房间的结构和布景的主要特色是使用当地的木料。每个住宅单元的框架结构和家具设施都由工厂的预制构件组装完成，整体外形尺寸为14.5m×3.5m，方便货车运输。住宅的整体设计风格、室内及外观的材质都由客户自己决定，有一份设计目录清单为客户提供各式各样的设计以供选择。这就为每个住宅单元都提供了量身制定的设计方案以及灵活的定价方案。由于采用模块化构件的搭建方式以及对木材的集约利用，隐形住宅的住宅单元可以被完全拆解，从而最大限度地减少空间占用。隐形住宅是一款价格合理、融创新性与流动性为一体的新型住宅产品，在当前日益严峻的住宅设计形势中，提供了一种突破性的住宅替代选择。这个独特的住宅创意，其核心要素在于简单的组装操作、富有吸引力的价格和住宅选址的自由性。相对于成本高昂、建造过程繁琐的传统住宅，这些元素显著体现了隐形住宅的主要价值。

Invisible House

Invisible House is a flexible housing unit which consists of a prefabricated wood structure designed for turnkey implementation at any designated site. Maximum flexibility and spatial quality are the key elements in its concept of development. The open layout is structured by a chimney and a wet cell creating three spatial units that provide for individual use and design. The structure and ambience of the rooms are characterised by the use of domestic woods. The mounting

定制过程
customization process

FACADE SELECTION
COLORED OR MIRRORED
01 STEP

02 STEP
INTERIOR SELECTION
FIR WOOD: TONAL RANGE

03 STEP
FURNITURE
OPTIONAL ELEMENTS

04 STEP
EXTERIOR
ADDITIONAL ELEMENTS

05 STEP
ON SITE
SETUP

framework and fitments of the housing unit are exclusively assembled from prefabricated elements at the factory. The overall dimensions are 14.50 x 3.50 meters, which provides for easy transportation by lorry. Design and texture of the interior design and facade can be determined by the client from various options listed in a design catalogue. This provides for tailor-made design options for the housing units as well as for flexible pricing options. Through modular element construction and the intensive use of wood, the housing units can be completely disassembled thus minimizing their environmental footprint. By combining innovation and mobility at a reasonable price, Invisible House is a product that offers a ground breaking alternative in an increasingly critical housing situation. Key factors in this unique proposition are its uncomplicated assembly, its attractive price and the free choice of location. Compared to the cost-intensive and bureaucratic construction of a conventional house, these elements represent the main assets of Invisible House.

项目名称：Invisible House
地点：Slovenia (movable)
建筑师：Delugan Meissl Associated Architects
项目经理：Gerhard Goelles
用途：living
总建筑面积：50m²
有效楼层面积：45m²
构造体量：160m³
设计时间：2013.2
施工时间：2013.5—2013.7
摄影师：©Christian Brandstätter (courtesy of the architect)

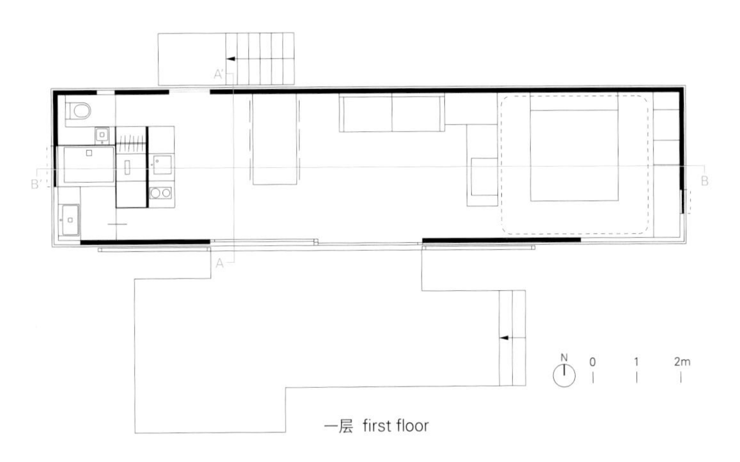

一层 first floor

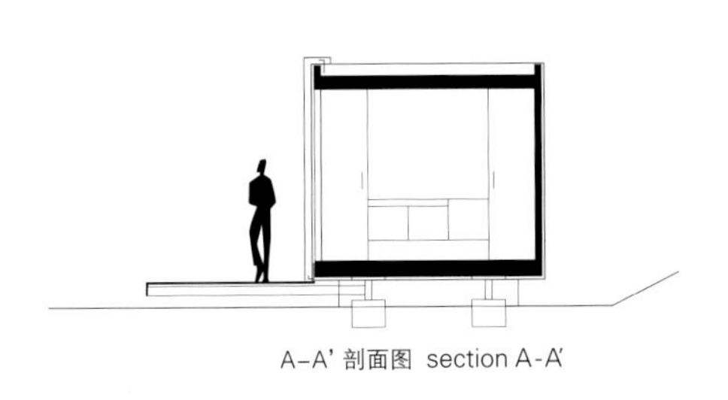
A-A' 剖面图 section A-A'

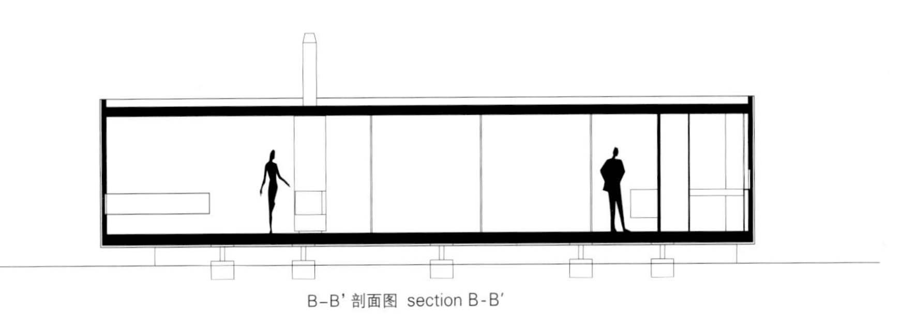
B-B' 剖面图 section B-B'

UFOgel小型住宅_Urlaubs Architektur

UFOgel小型住宅是一座非凡的建筑，装修得很舒适，几乎完全是用落叶松木打造的。它的名字也是根据它的外形而来的。有时候它看起来像一只始祖鸟，有时候看起来像外太空的建筑，总之看起来非常特别。它无疑不是传统的建筑，相反，它更像是一个雕塑。当你走近它，就知道它确实是个空间奇迹。

这座住宅的建筑面积为45m²，可满足多种功能，包括舒适的厨房配备、卫生间和浴室等。大面积的全景窗户将自然风景引入室内，而弯曲的传统木质覆盖板营造了温暖、安全的感觉。落叶松木的材质类似于瑞士五针松，标志着阿尔卑斯山地区建筑的传统。对于它的新主人来说，UFOgel小型住宅的每一根木头都是新鲜的，也可以说是金黄色的，随着时光的流逝它会变成灰白色。这座小屋与环境融合，成为大自然的一部分，就像东蒂罗尔的木舍和山间房舍一样。

拥有UFOgel小型住宅就像拥有了一个个性建筑。它是一座活的建筑，随着时光的流逝它逐渐改变、成熟，获得尊严和美丽。只有很少的建筑才能获此殊荣，即能被称为活的建筑。拥有独特的个性和品质意味着建筑很有价值。

无论使用者是坐着还是躺着，淋浴或者享用晚餐，在这个度假屋里永远可以看到这两样东西——木材和天空。

UFOgel小型住宅拥有具有保护性、紧凑、节省空间和防风雨的外形。它的设计源于对空间价值的了解，对严寒的冬天和暴风雨天气的了解，以及对木材质量尤其是落叶松品质的了如指掌。

Biwak露营民舍和山间小屋是UFOgel小型住宅的始祖，UFOgel小型住宅非常舒适，但是它从不否定其来源于高山地区以及对提供住宿的特殊需求。

UFOgel小型住宅同时提供了开放性和安全性。宽阔巨大的全景天窗将景色纳入屋内，使人产生了身处室外的感觉。对比之下，圆形的屋顶提供了舒适的木材质感。这使UFOgel小型住宅成为一个温暖神秘的居所，在这里睡眠非常舒适而且有种返璞归真的感觉。

作为两层住宅，UFOgel小型住宅适应了居民的需求。它既可以作为船库、阿尔卑斯山的小屋，又可以作为度假屋和展示建筑。它时而开放张扬，充满魅力，时而是非常隐秘的避难所。

UFOgel

The UFOgel is an extraordinary building, furnished comfortably and almost completely manufactured with larch wood. Its shape gives this building its name. Sometimes it looks like an archaeopteryx, sometimes like a building from outer space, but always like something special. It is definitely not a conventional building but rather a sculpture, which turns out to be a spatial miracle when you enter it.

45m² living space, furnished comfortably with a kitchen unit, bathroom and designer shower, can be used multifunctional. Big panorama windows bring the nature inside and the wooden body of the vaulted building, which is covered with traditional shingles, evokes a feeling of security and warmth. The larch wood marks – similar to the arolla pine – the heritage of the alpine architecture. Every UFOgel comes fresh, you can also say blond, to its new owner and becomes grey with

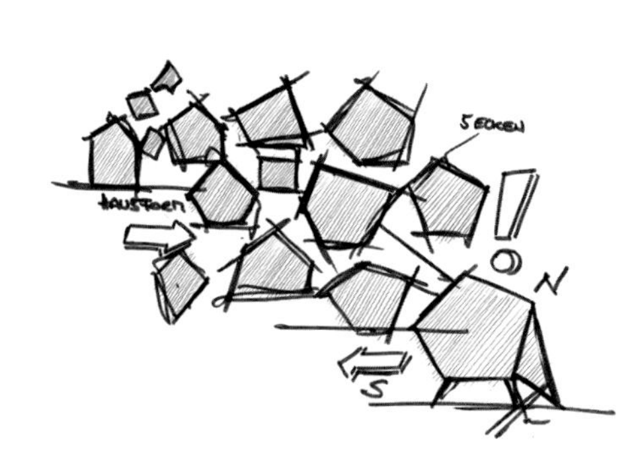

项目名称：UFOgel
地点：Nussdorf-Debant, Tyrol, Austria
建筑师：Urlaubs Architektur
项目团队：Peter Jungmann, Lukas Jungmann
用地面积：780m²
总建筑面积：45m²
施工造价：EUR 190,000
竣工时间：2012
摄影师：©Lukas Jungmann (courtesy of UFOgel)

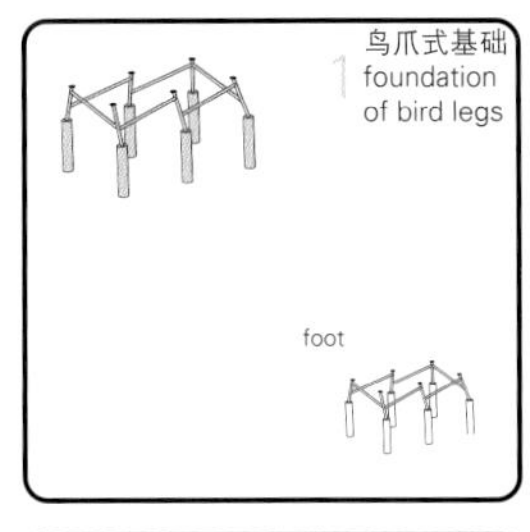

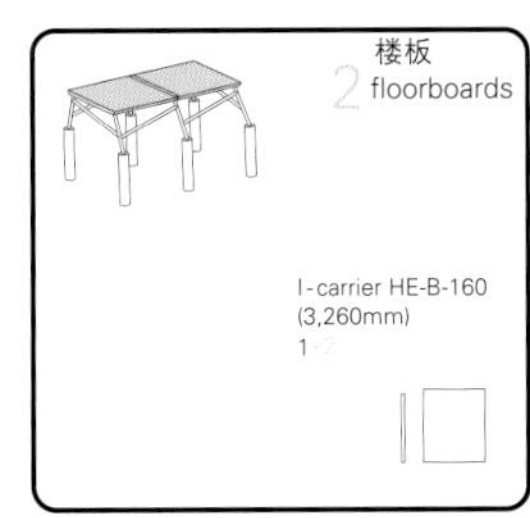

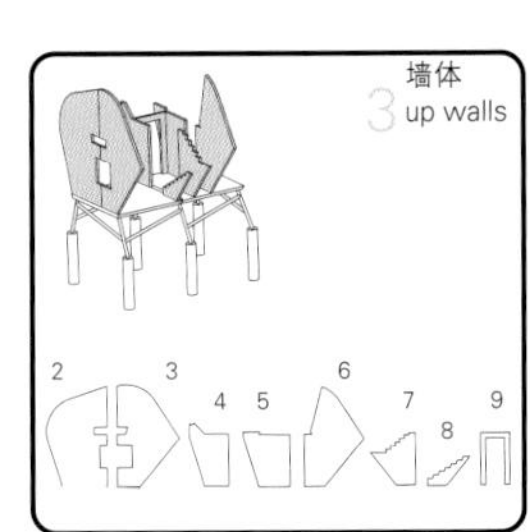

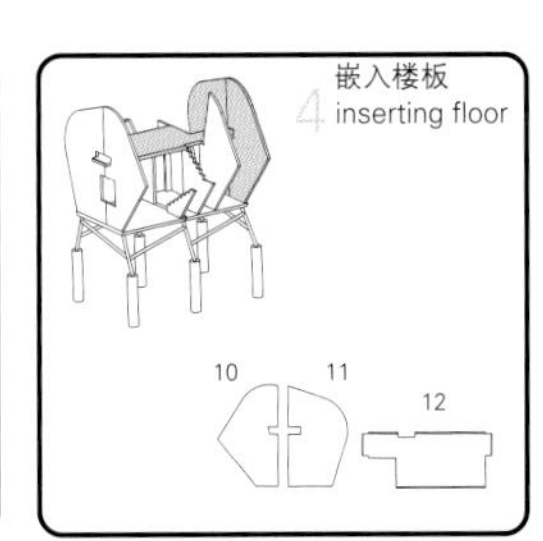

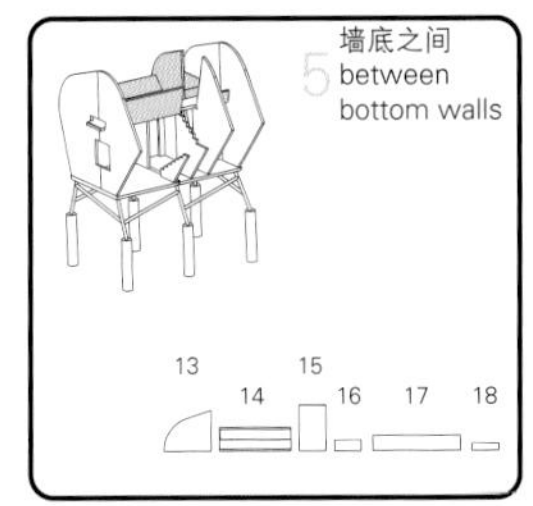

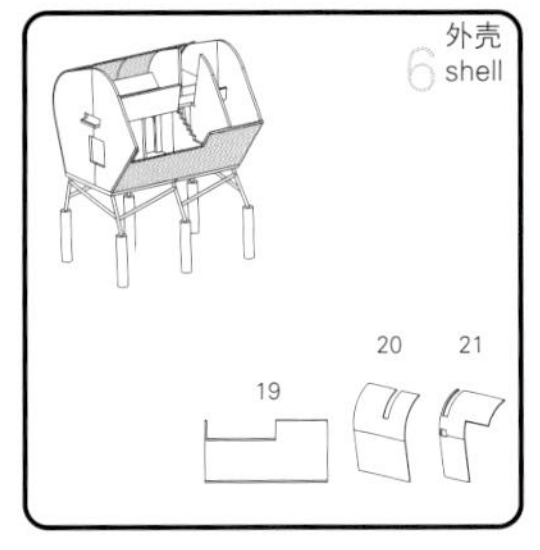

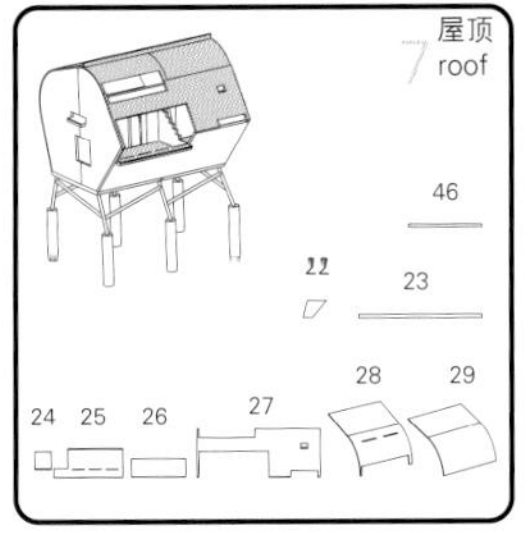

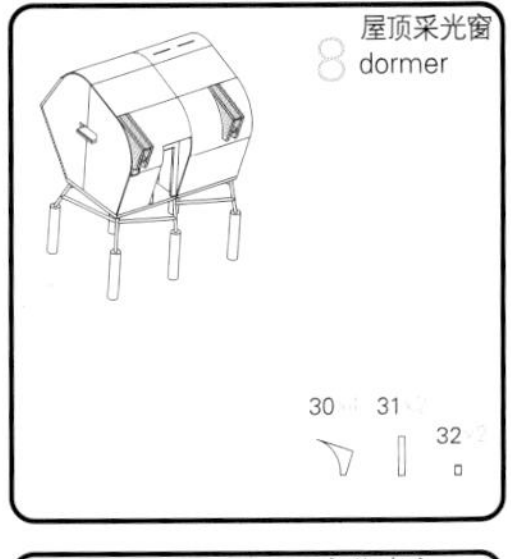

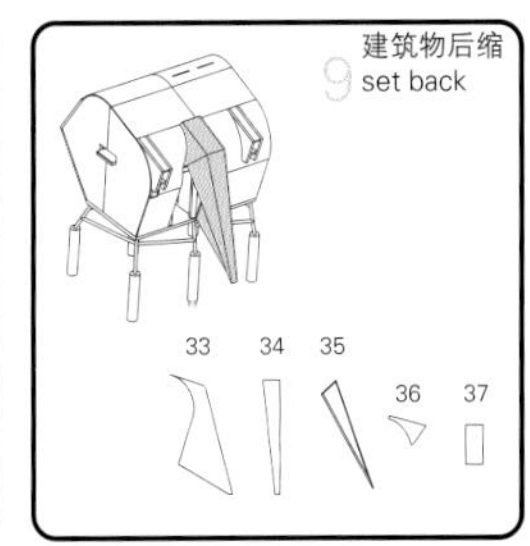

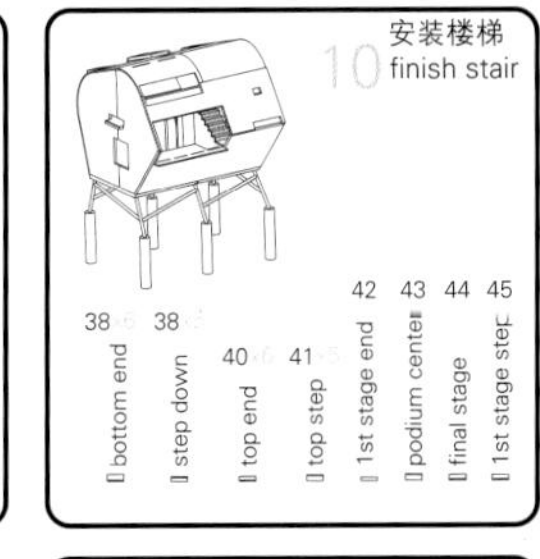

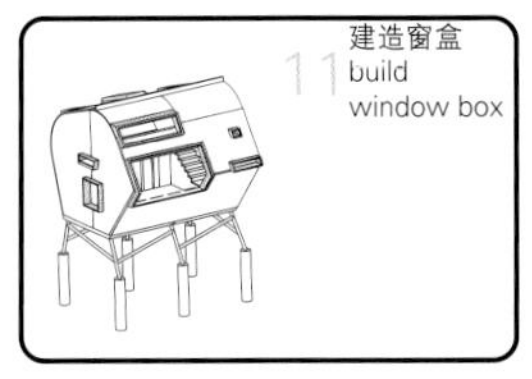

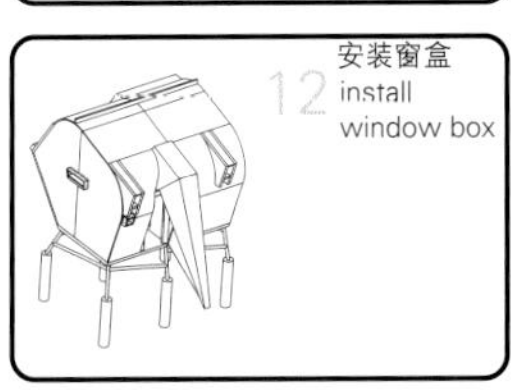

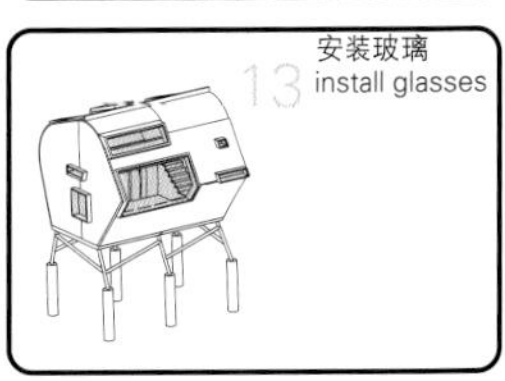

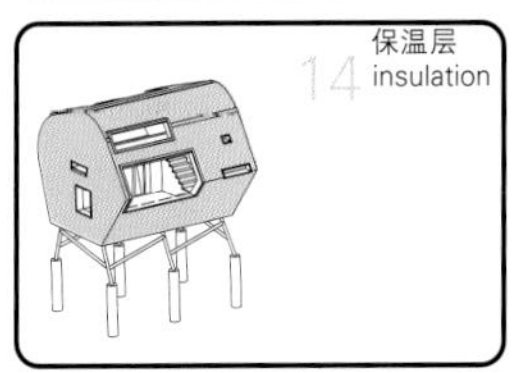

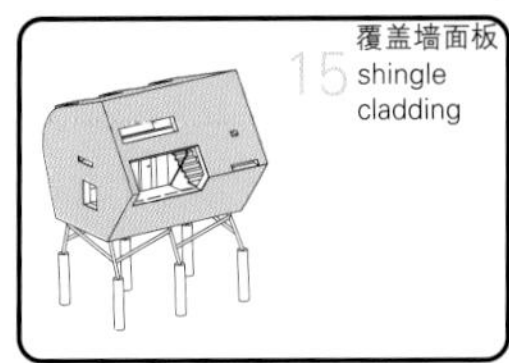

age. It melts together with its environment, steps back and becomes a unit with nature, like the wooden farmhouses and mountain hugs in eastern Tirol.

To own a UFOgel means to own a building with personality. A living building, which changes, matures and gets dignity and beauty with age. There are only a few buildings, about whom you can claim this. Having character and quality means to be valuable.

While sitting, lying, showering or dining in the UFOgel two things are always present: the wood and the heaven.

The UFOgel has a protective, compact, space-saving and weather resistant shape. Its design is a result of the knowledge of the value of space, of the weather conditions during rough winters and storms and of the qualities of the material wood, especially of larch.

Biwak and mountain hugs are the ancestors of the UFOgel which can be very comfortable but does never deny to be from the high mountains with their special requirements for an accommodation.

The UFOgel offers openness and security at the same time. The big panorama window brings in the landscape and evokes the feeling to be outside. In comparison, the round shell offers a pleasant and wooden coziness. It makes the UFOgel a shelter with all the warmth and quaintness, with healthy sleep and the feeling to return to the origins of living.

On two living levels, the UFOgel adjusts to the needs of its inhabitants. It can be boathouse or alpine hut, holiday flat or exhibition pavilion. Sometimes open and inviting, sometimes a very private refuge.

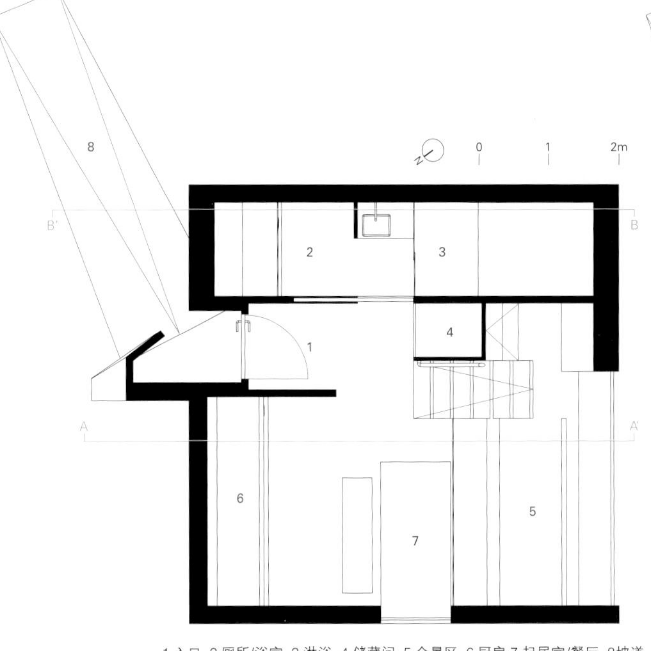

1 入口 2 厕所/浴室 3 淋浴 4 储藏间 5 全景区 6 厨房 7 起居室/餐厅 8坡道
1. entrance 2. toilet/bath 3. shower 4. storage space
5. panorama area 6. kitchen 7. living room/dining room 8. ramp
一层 first floor

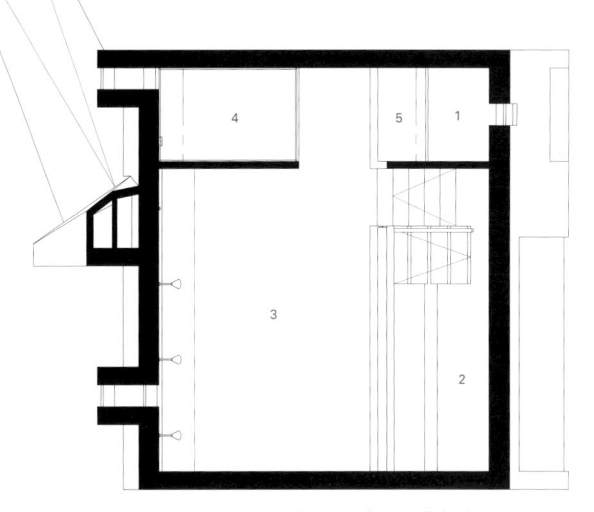

1 淋浴 2 全景区 3 起居室 4 卧室 5 开放式衣柜
1. shower 2. panorama area 3. living room
4. bedroom 5. open garderobe
二层 second floor

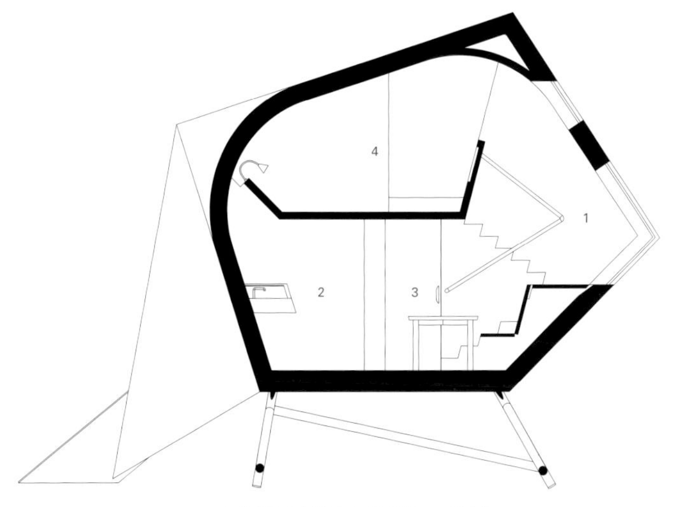

1 全景区 2 厨房 3 起居室/餐厅 4 起居室
1. panorama area 2. kitchen 3. living room/dining room 4. living room
A-A' 剖面图 section A-A'

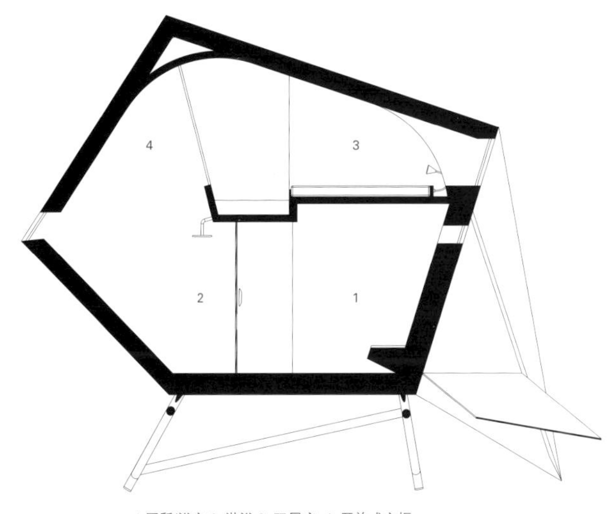

1 厕所/浴室 2 淋浴 3 双层床 4 开放式衣柜
1. toilet/bath 2. shower 3. double bed 4. open garderobe
B-B' 剖面图 section B-B'

帐篷水疗中心_a21 Studio

这座小型的帐篷水疗中心坐落在越南一个温泉矿物度假村，它悄然地栖息在岩石山半山腰延展的露台上，直面河水。山的另一边熙熙攘攘，这儿似乎是躲开那纷扰尘世的绝佳场所，令人静享健康的生活方式，如泥浴和按摩等。然而，场地最大的局限之一是全年沉浸在越南西部地区的酷热气候之中，因此此处多年来一直是休耕地。

首先茅草覆盖的倾斜屋顶被认为是阻隔日晒的好方法，同时还能很好地与周边的景观相融合。倾斜的一边朝西，大量的热量将被阻隔在建筑之外。由此形成屋顶仿佛从地面弹出来的效果，赋予帐篷一种独特的外形，屋顶不再只是一般地遮风挡雨防日晒了，而是成为建筑的主要结构。具体来说，100mm×250mm的横梁使得结构跨度达到了11m，从2.3m的半空处抬高，游客得以欣赏到美丽的河景和市中心的景观。在这个木结构之上，三层20mm厚的木板相结合，使天花板显得格外美观，所有的横梁、防水膜、30mm厚的椰子叶也由此各自相连。当地建筑技法和榫卯接合实现了各个部分之间的相互连接。

游客能够感受到浓郁的本土风情以及一种和谐的氛围，这得益于各种本土材料的使用和施工手法，包括就地取材的叠石和椰子叶、岩石、钢筋和各种瓷砖。在帐篷水疗中心内部，空间被分割为两层。上层是私人客房，笼罩着木质框架和彩色玻璃，其中木材来自40多年前停产的旧式木质框架。同时，安装在上空空间中央的网让游客感受到了一种异域情调，人们可以在半空中尽情玩耍。下层空间有个无边矿物泳池，配套旧式木质家具，成为水疗中心的休憩场所，可以俯瞰其下方的河景。

总之，帐篷水疗中心连同别具一格的屋顶结构和相宜的材料选取，复兴了这处被遗弃之地，使其充满阳光和微风。项目的实现离不开当地工人纯熟的施工技法，它因而成为周边环境的典范之作。

The Tent

Located inside an operating hot spring and mineral resort, the Tent is a small spa perched in the folds of halfway terrace up to a rock hill, facing to the river. This seems to be the good place for hiding away from the eventful area on the other side of the hill to enjoy healthy activities such as mud bathing and massage services. However, one of the biggest constraints of the site is that it gets intense heat from the west throughout the year which makes it fallow for years.

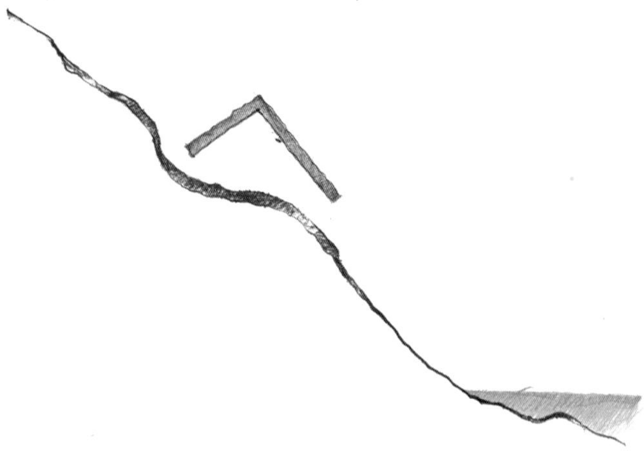

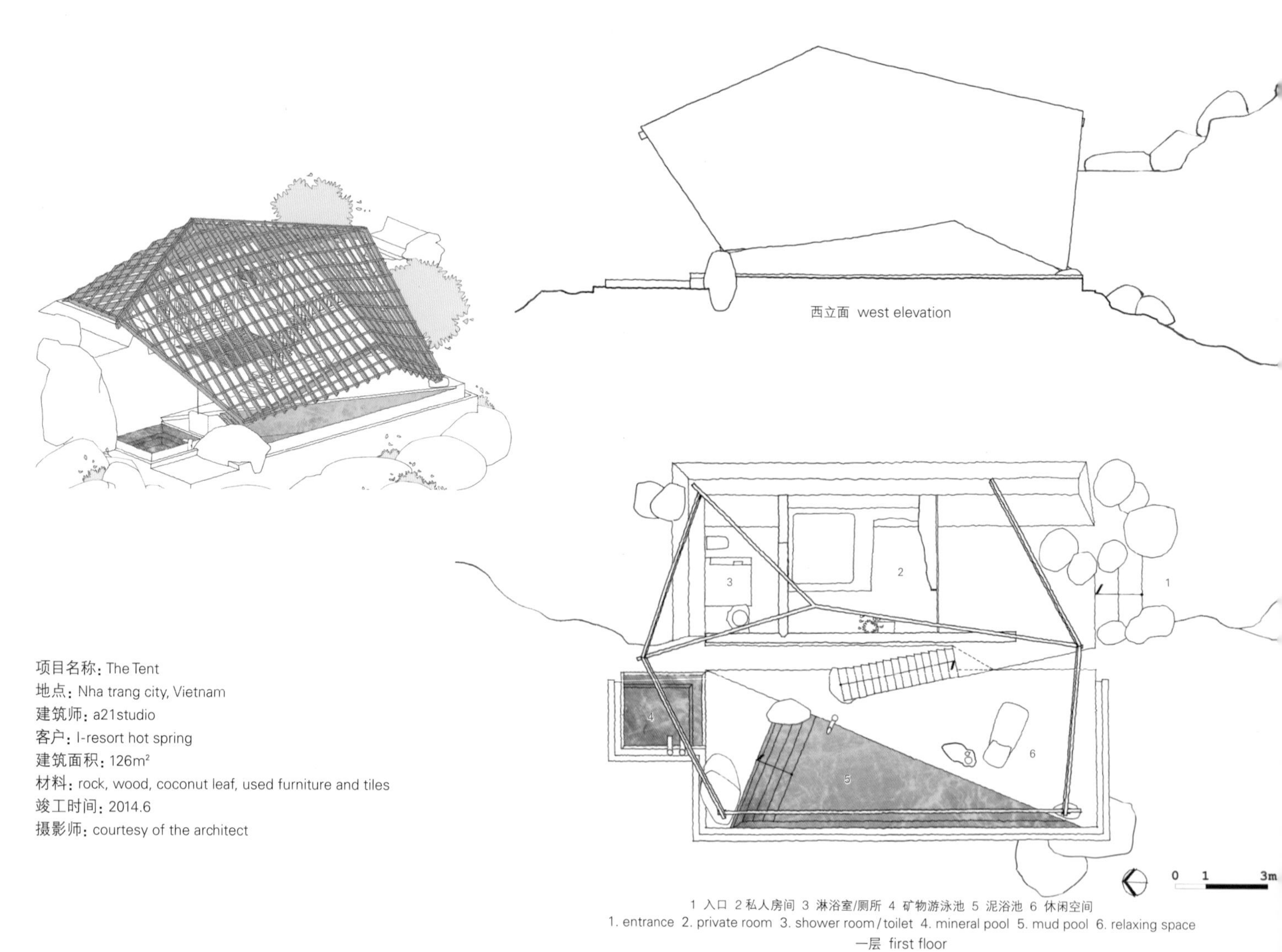

西立面 west elevation

1 入口 2 私人房间 3 淋浴室/厕所 4 矿物游泳池 5 泥浴池 6 休闲空间
1. entrance 2. private room 3. shower room/toilet 4. mineral pool 5. mud pool 6. relaxing space
一层 first floor

项目名称：The Tent
地点：Nha trang city, Vietnam
建筑师：a21studio
客户：I-resort hot spring
建筑面积：126m²
材料：rock, wood, coconut leaf, used furniture and tiles
竣工时间：2014.6
摄影师：courtesy of the architect

Firstly, a steep roof with thick thatch is considered to be a good solution to prevent the sunlight as well as a good blend with surrounding landscape. By placing the slope side to the west, a huge amount of heat is trapped outside the building. Furthermore, the fact that the roof props on the ground also gives a distinctive look to the Tent since the roof is no longer used as a cover from either rain or sun as normal way but also becomes the main structure of the building. In detail, the main 100x250mm beams help the structure spanning 11m and elevating 2.3m in the half way to allow guest enjoy

a good view to the river and city center. Above this wooden structure is a combination of 3 layers, 20mm thick wood panels which gives an aesthetic look to ceiling and links all beams together, waterproof membrane and 30mm coconut leaves, respectively. All of these parts are connected by using indigenous building technique, mortise and tenon joints. Together with the construction method, local materials, such as dry-stacked stones and coconut leaves quarried right on the site, as well as rocks, reinforced steels and different tiles are adopted to bring a harmonic and native feeling for the guests. Inside the tent, the space is divided into two levels. On the upper floor, a private bedroom that is covered by wood frames and color glasses, collected from old wooden frames with more than 40 years that no longer produced. Besides, a net, installed in the middle of the void, adds an exotic feeling to the guests, who enjoy playing in the mid-air. Down to the below space, an infinitive mineral pool with old wooden furniture is a relaxing space for the Tent that look to the river below.

In conclusion, the Tent with an extraordinary roof structure and harmoniously material combination has revitalized an abandoned land full of sun and wind. This either honors skill-ful construction techniques of local workers or becomes a typical example for its surrounding area.

滨海绍森德海边小屋_Pedder & Scampton Architects

这片位于英国舒伯里内斯的基地是海岸线的延伸部分，恰好在绍森德的一个角落里。这里有着广阔的天空和大型木质防波堤以及一条冷战时期的分界线、大型船只，远处还可眺望到发电厂。海滩次第为卵石、贝壳、破碎的砖头，以及港湾的淤泥。海边小屋直到20世纪80年代都坐落于此，多年间持续地排列在滨海绍森德海边的大部分地区。

我们在整体的外形下赋予了这些海滩小屋强烈的节奏感和轮廓感，它是一种整体模式下的多样化和个性化，一种强大而简单的造型，也是大型景观之内的私人空间。

我们的设计提案中的海滩小屋十分独特，从后面的人行道和小屋上方可以看到其起伏的屋顶轮廓。小屋之间的空隙使得面向大海的视野更加明显，而绿色屋顶则反映了海滩和陆地之间的绿化范围。小屋面向南边，不对称的平面设计使得前方形成了大量座位区，可俯视美景，而小屋后侧为一处狭窄的工作和库存区域。小屋的朝向和造型意味着每间小屋前方都有一个面向海滩的私人外部空间，而且两面的风景都不会因为小屋的存在而被忽视。

海滩小屋的墙由可循环的木材制成，形成一个框架，内部布满夹层，外部是三层聚碳酸酯墙板。墙体里面填充卵石、玻璃碎片、贝壳，通过分层铺设的方式创造了每个小屋独特的样式，同时提供了保温性、私密性和安全性。屋檐上透明的聚碳酸酯板让阳光照射到屋里。从上面看屋顶就是简单的倾斜木托盘覆盖着海洋类天属植物。小屋包括被涂成不同颜色的全高胶合板。胶合板的一部分组成固定的门，另一部分是半固定的。门可以部分或者全开来适应天气变化和居住者的需求。固定的门板在小屋内部创造出有遮挡的变化的空间，门板外部用工业化的字体漆出小屋的数字。巨大的内部夹层空间使每个小屋的主人都能轻而易举地定做自己的内部空间。

Southend-on-Sea Beach Huts

The site at Shoeburyness is an untamed stretch of coastline just around the corner from Southend. It is characterized by big skies and timber groynes with distant views of a cold war barrier, huge ships, and power plants. The beach grades from pebbles, shells and broken bricks down to estuarial mud. Beach Huts were positioned on the site until the 1980s and

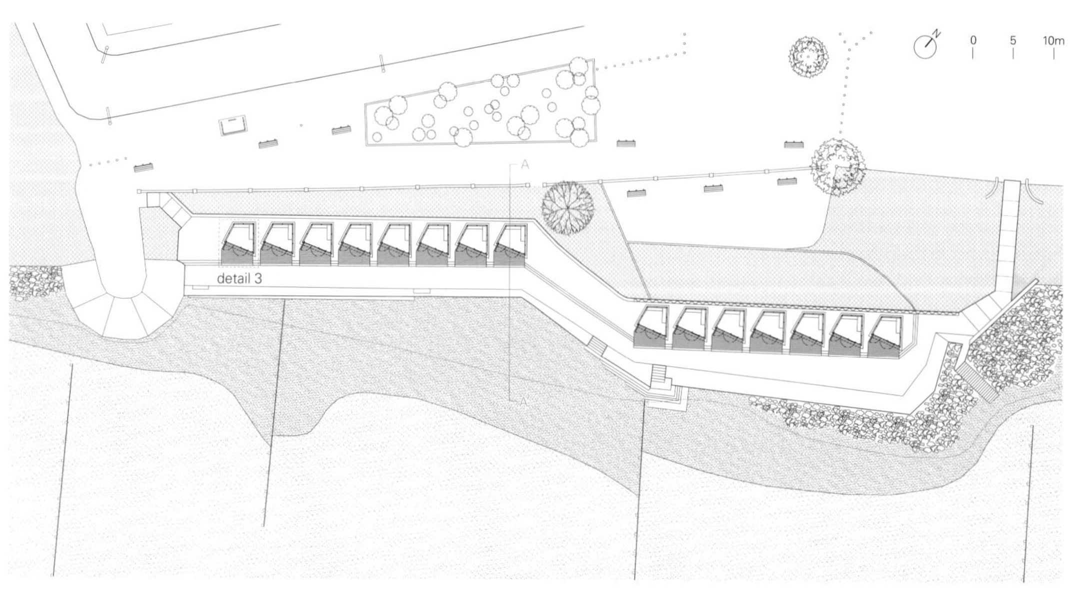

2
4
5
7
8

©Peter Cook(courtesy of the architect)

墙体安装
wall build up

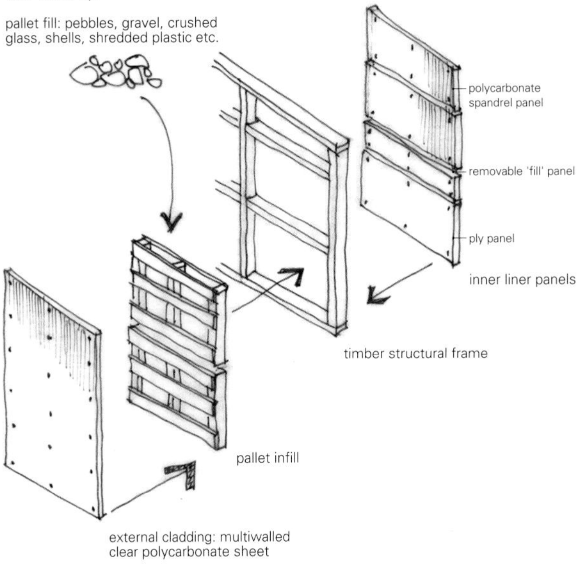

pallet fill: pebbles, gravel, crushed glass, shells, shredded plastic etc.

polycarbonate spandrel panel

removable 'fill' panel

ply panel

inner liner panels

timber structural frame

pallet infill

external cladding: multiwalled clear polycarbonate sheet

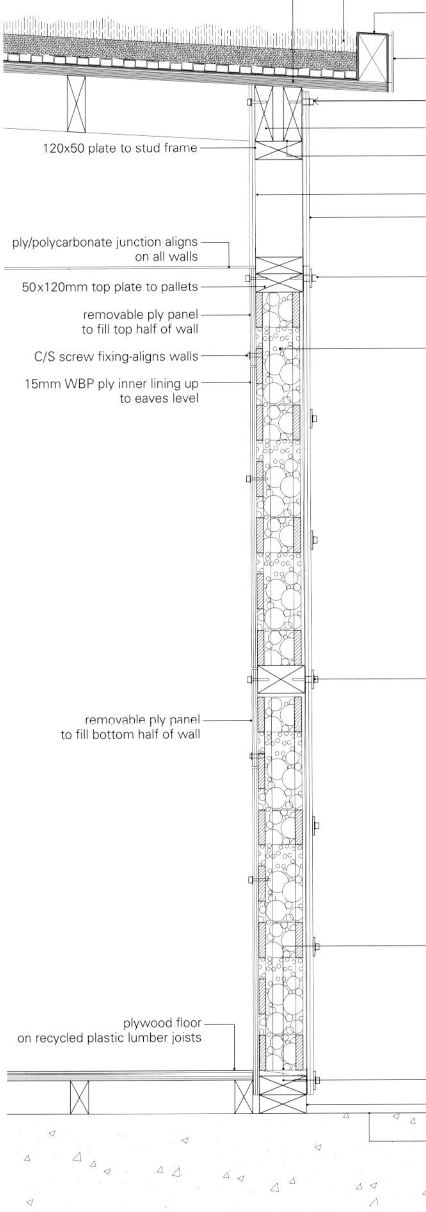

roof deck 2 layers 15mm plywood, glued and screwed together

roof build up: sedum mat on growing medium on filter fleece on drainage/reservoir boards on waterproofing
in timber tray as specialist requirements

75x100mm deep perimeter section

aluminium flashing

screw fixing with 25mm DIA. washer

full depth noggin

120x50 plate to stud frame

50x150 end plate to joists fixed to wall panel

16mm triple wall polycarbonate sheet internally

16mm triple wall polycarbonate sheet spaced 5mm off the framing/pallets to allow for drainage and ventilation

ply/polycarbonate junction aligns on all walls

50x120mm top plate to pallets

5.5mm DIA. 55mm SS security screw with 50mm DIA. SS washer with EPDM backer. 32mm DIA. 5mm thick rigid spacer behind polycarbonate

removable ply panel to fill top half of wall

C/S screw fixing-aligns walls

1,200mm(w)x1,000mm(h) reconditioned timber pallets if required, no wall fill

15mm WBP ply inner lining up to eaves level

fixings align on walls timber bearer

removable ply panel to fill bottom half of wall

wall clapping recycled hardwood from southend pier

bottom row of pallets mounted on 15mm (h)x 50mm(w) intermittent packers
DPC fixed over packers, under pallets and wrapped up inner face and down sole plate behind packers to create drainage/ventilation path below pallets applies to walls large fill pieces
at low level to create drainage paths.

plywood floor on recycled plastic lumber joists

120x50 timber base plate

50x150 recycled plastic lumber sole plate

existing concrete platform

详图1 detail 1

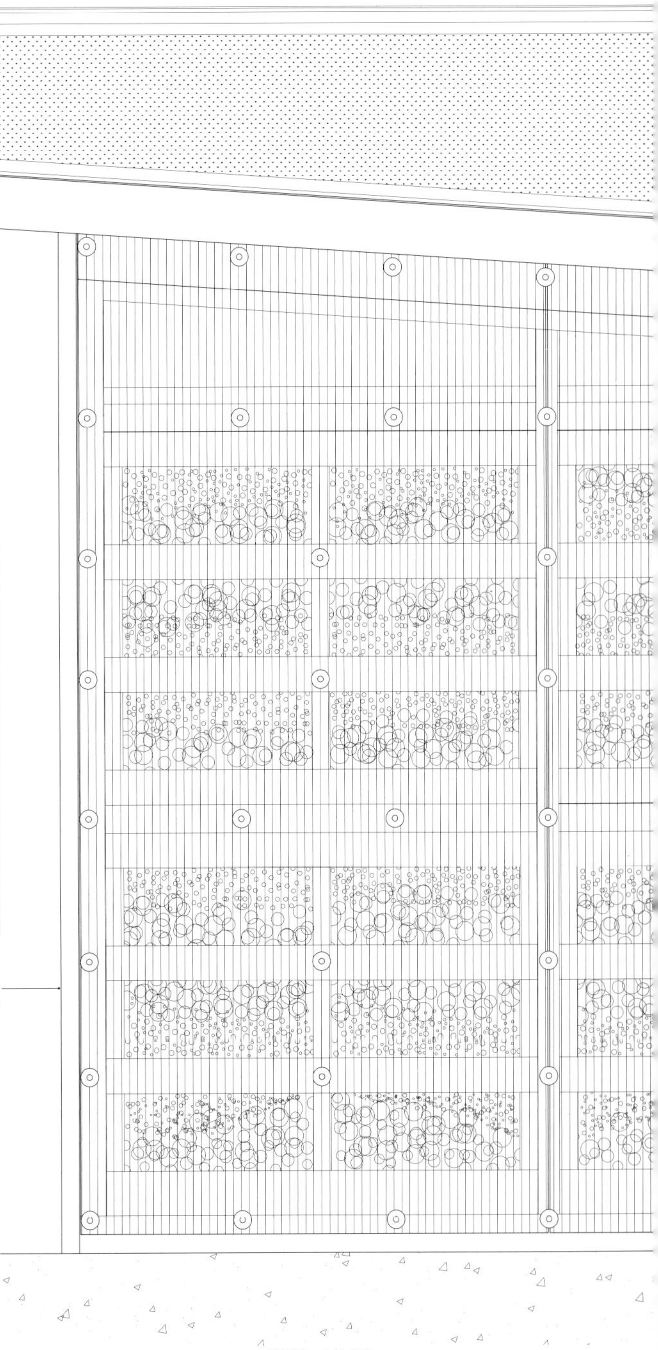

详图2 detail 2

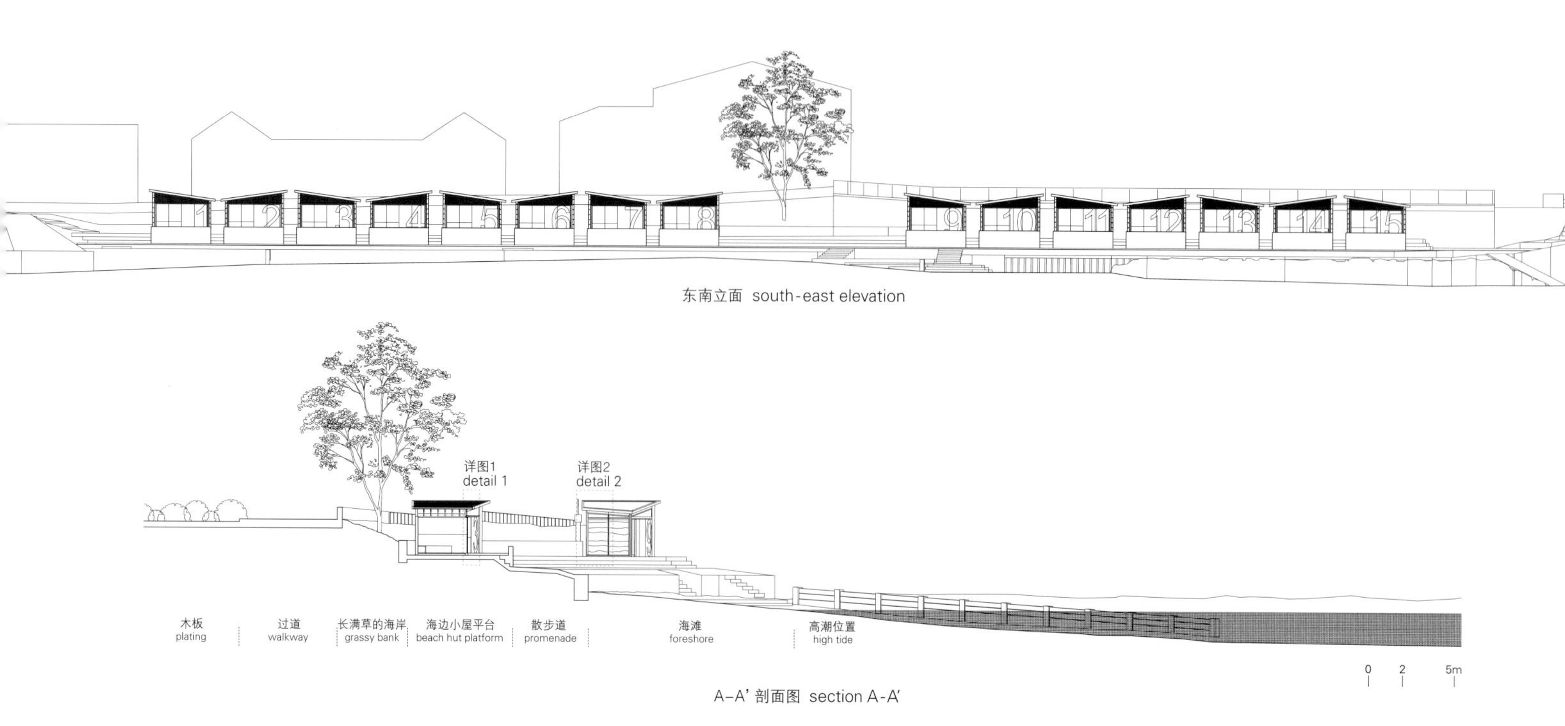

东南立面 south-east elevation

A-A' 剖面图 section A-A'

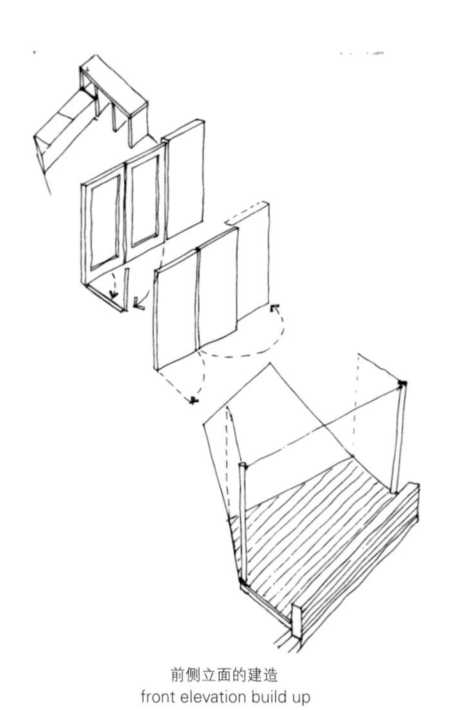

前侧立面的建造
front elevation build up

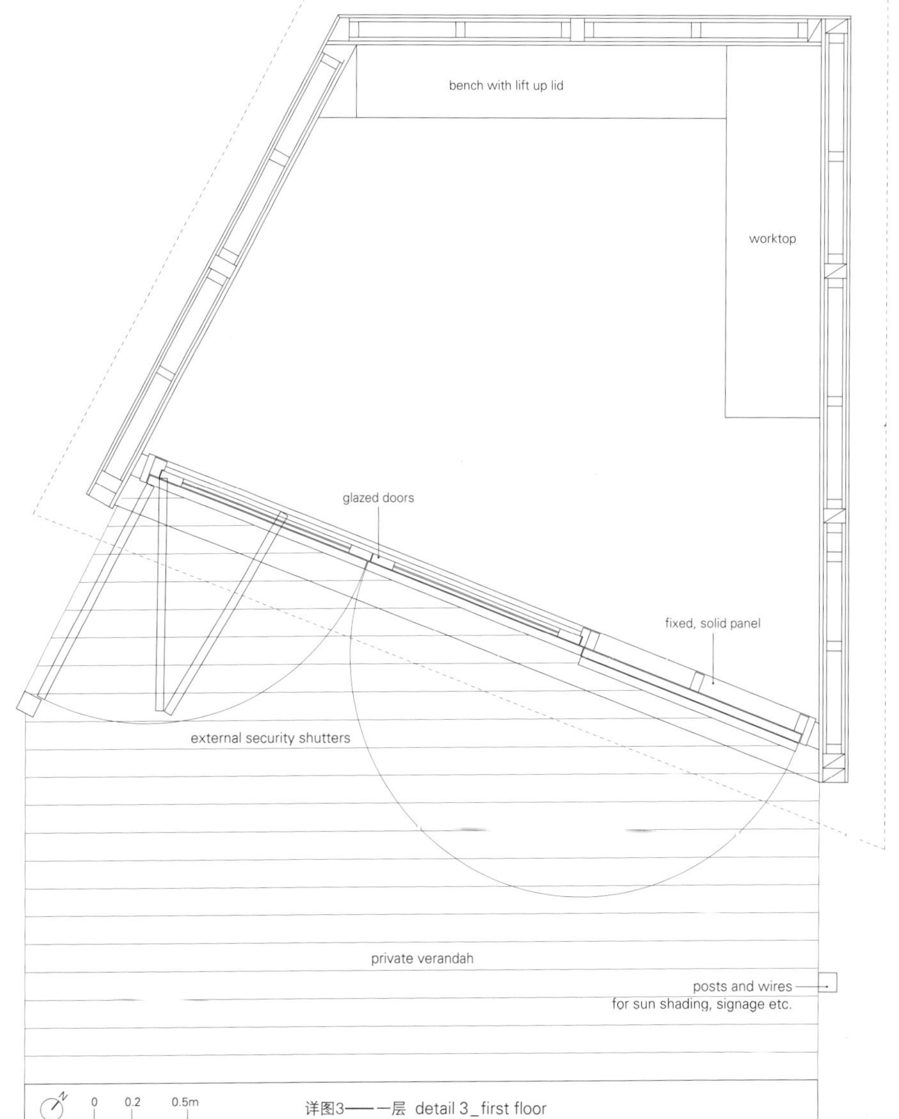

详图3——一层 detail 3_first floor

项目名称：Southend-on-Sea Beach Huts
地点：East Beach, Shoeburyness, Southend-on-Sea, United Kingdom
建筑师：Helen Pedder, Gillian Scampton
结构工程师：Geoff Morrow, Ed Hollis (StructureMode)
工料测量师：MPA Construction Consultants
承包商：Carmelcrest Ltd
资金：Southend Borough Council
用地面积：295m²
建筑面积：24m²(each beach hut)
委托时间：2012.12
设计时间：2013.3
施工时间：2013.6—2013.8
客户：Southend on Sea Borough Council
摄影师：©Simon Kennedy (courtesy of the architect)

continue to line much of the Southend sea front.

We took the defining characteristics of beach huts to be a strong rhythm and profile, a sense of variety and individuality within the overall pattern, a robust and simple form, and the creation of an intimate space within a big landscape.

Our proposal creates distinctive beach huts with an undulating roof profile that can be seen from the promenade behind and above the site. Gaps between the huts allow views out to the sea and the planted roofs reflect the existing green margins between the beach and the land. The huts are oriented to the south with an asymmetrical plan which gives plenty of sitting space at the front overlooking the view and a narrower working and storage space to the rear. The orientation and form of the huts means that each hut has a private external space in front of it, open to the beach and the view but not overlooked by the huts to either side.

The beach hut walls are formed from recycled timber pallets, bolted into a frame and lined internally with ply and externally with triple wall polycarbonate sheeting. The wall cavities will then be filled with pebbles, crushed glass, shells etc. laid in strata to create an individual pattern for each hut as well as providing thermal mass, privacy and security. Above eaves level a clear polycarbonate lining allows sunlight into the huts. The roofs, seen from above, are simple canted timber trays, filled with sea hardened sedum. The sea elevation consists of full height plywood panels painted in a range of colors, one half divided into stable doors and one half fixed. The doors can be half or fully opened as suits the weather and the occupants, while the fixed panels create screened changing space inside the huts and are painted with industrial scaled hut numbers on the outside. The robust internal ply linings make it easy for each hut owner to customize their own space.

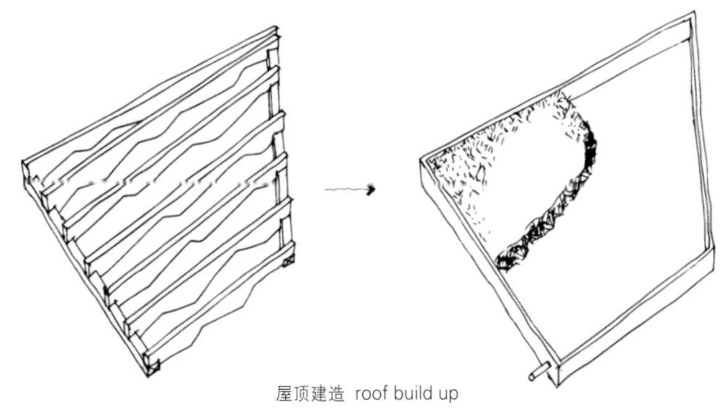

屋顶建造 roof build up

里多克旅馆_Gabriel Rudolphy + Alejandro Soffia

在我们的社会中，人们习惯性地认为建筑是奢侈品，同时认为奢侈品是有价值而且稀少的。我们同意这个定义的前一点，但是关于第二点，作为建筑师我们也许并没有足够的坚持去建设一个品质更好的环境，无论是室内还是室外。

如果我们把奢侈品定义为布鲁诺·莫拿利的作品，那么我们就意识到了，一般来说高端的建筑产品与昂贵材料的不恰当使用联系在一起。在一定程度上，这就产生了不平等，即优秀设计的益处只有买得起这个建筑产品的人们才能享用。同时为了让建筑更加接近大部分人的生活，我们一直在致力于改善设计质量与成本之间的关系。为了达成这个目的，我们假设合理的构造将定义设计法则，因此我们会选择设计此类建筑项目，即将建筑系统的材料条件最优化，用低成本和技术简单的材料来建造房屋。利用建筑材料的多元化属性，我们得以在建造的时候充分利用材料进而减少时间和成本。

在里多克旅馆项目中，我们的概念关注优质的低成本设计，这一点与客户的需求及其预算之间取得了平衡。就像经常在智利所发生的，设计愿望很美好，现实预算却很紧张。而对于那些建筑产业的局外人来讲，他们不熟悉建筑成本，每个人都希望他们用毕生工作所得进行的投资能够满足他们所有的预期。所以，为了减少成本，也因为项目坐落在乡村地区，我们开始选择当地的技术和劳动力来建造这个项目。换句话说，我们联系了擅长这个地区建筑类型的"手艺人"——简单的锯切松木是这个地区有代表性的建筑材料——然后我们在最普遍的当地建筑方法的基础上建立了施工技术目录。这个决定让我们减少了运输成本和施工利润率。在我们掌握了该地的建筑系统之后，第二个关键要素就是最大化地利用其最常见的木材尺寸。这个尺寸是由适合整个项目的不同体积的宽度决定的。这个决定本身不会节省成本，但是最大限度地调整了空间，为建筑增加了附加价值，大部分的建筑构件（2.4m）都不会达到这样的效果。

里多克旅馆项目坐落于同名海滩的最北端。20世纪70年代发展成为瓦尔帕莱索学校的"开放城市"。整个建筑格局被分成五个独立的部分，三个部分是两层的耳房，一个部分是带有公共空间的服务区，一个

部分是业主的公寓。每个部分都分布在平面布局上，以求获得独立性以及最佳的观景视野。

Ritoque Hostel

In our society, we customarily perceive architecture as a luxury and understands luxury as something valuable and at the same time scarce. We agree with the first component of the definition, but with respect to the second point, we as architects have probably not been persistent enough to build a better quality environment, nor inside or outside the centers. Now if we define luxury as Bruno Munari does, we realize that, in general, high-level architectural production is associated with improper use of costly materials which produce an equality gap to the extent that the benefits of good design can only be taken advantage of by those that can afford it. Likewise, and in order to bring design closer to a larger percentage of the population, we have been working for some time to improve the relation between design quality and cost. To accomplish this, we started from the hypothesis that rational construction should define the laws of design and, therefore, we have opted to design projects that optimize the material conditions of building systems that are low-cost and technologically simple. Capitalizing on the dimensional properties of construction materials allows to take advantage of materials so as to reduce the times and costs involved in a work of architecture.

项目名称：Ritoque Hostel
地点：Playa de Ritoque s/n, Quintero,
Región de Valparaíso, Chile
建筑师：Gabriel Rudolphy, Alejandro Soffia
结构工程师：José Manuel Morales, Gabriel Rudolphy
总承包商：Juan Tapia, Francisco Tapia, Diego Arenas
业主：Diego Arenas, Dayenú Vencilla
用地面积：402m² / 建筑面积：130m²
有效楼层面积：140m² / 造价：USD 650/m²
设计时间：2012 / 竣工时间：2014
摄影师：
Courtesy of the architect-p.50, p.53
©Pablo Casals-Aguirre (courtesy of the architect)-p.48, p.49, p.51

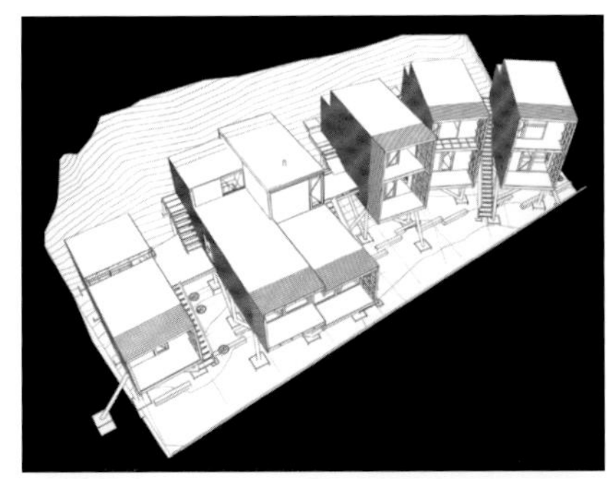

屋顶 roof

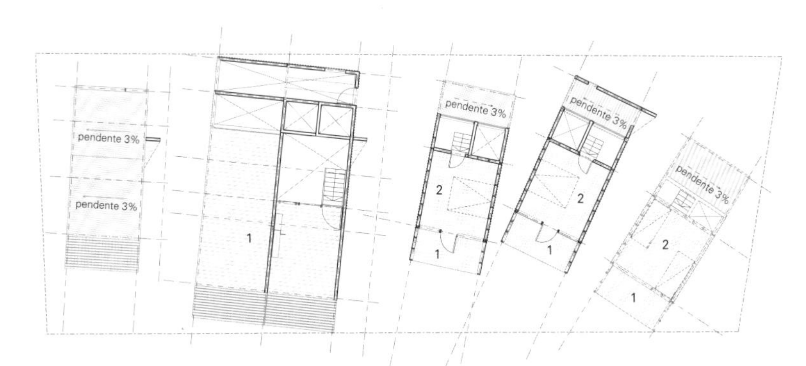

1 露台 2 卧室 1. terrace 2. bedroom
二层 level +240

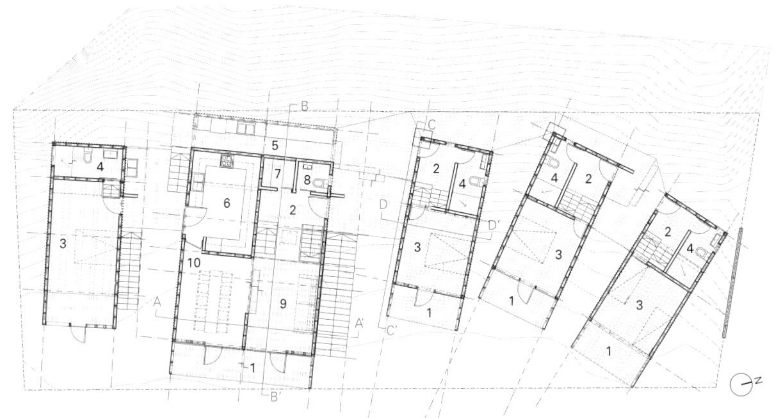

1 露台 2 门厅 3 卧室 4 浴室 5 洗衣房 6 厨房 7 设备间 8 厕所 9 起居室 10 餐厅
1. terrace 2. hall 3. bedroom 4. bathroom 5. laundry 6. kitchen 7. technical 8. wc 9. living room 10. dining room
一层 level 0

地下层 level -300

In the case of Ritoque Hostel, our conceptual concerns for good, low-cost design coincided with the equilibrium point between the client's wishes and his budget. As often happens in Chile, wishes were lofty and budget tight. While it's normal for industry outsiders to be unfamiliar with building costs, everyone expects the investment of their life's work to meet all of the expectations their imagination can come up with. So, in order to reduce costs and because the project was located in a rural area, we began the project by selecting local technologies and labour. In other words, we contacted the "tradesman" responsible for the construction typology in the area – simple architecture of sawn pine timber – and we created a catalogue of construction techniques based on the most common local solutions. This decision allowed us to reduce transportation costs and construction profit margins.

Once we were well versed on the construction system, the second key aspect consisted of optimizing the dimensions of the most common length of wood. This measurement is defined as the width of the way of the different volumes that compromise the project. This decision in itself produced no savings but added value of adjusting spaces to a greater measurement than that of most constructive elements (2.4 m) did. Ritoque Hostel is located in the far north end of the beach of the same name, were in the 1970's was developed the "Open City" of the Valparaiso School. The layout was divided into 5 independent volumes: three for two-story bays, one for service areas with common spaces and one apartment for the owner. They are each distributed in the floor plan to achieve independence between them, and the best orientation towards the landscape. Gabriel Rudolphy + Alejandro Soffia

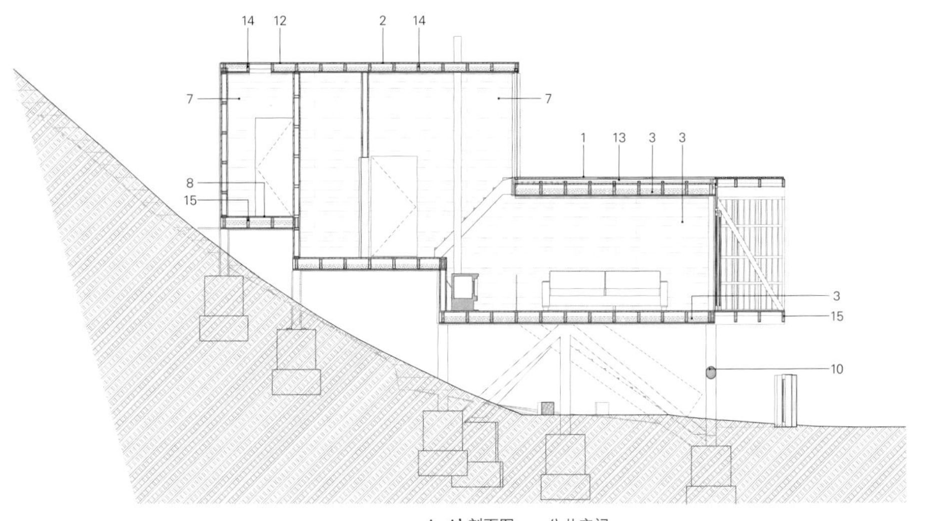

A-A' 剖面图——公共空间
section A-A'_common space

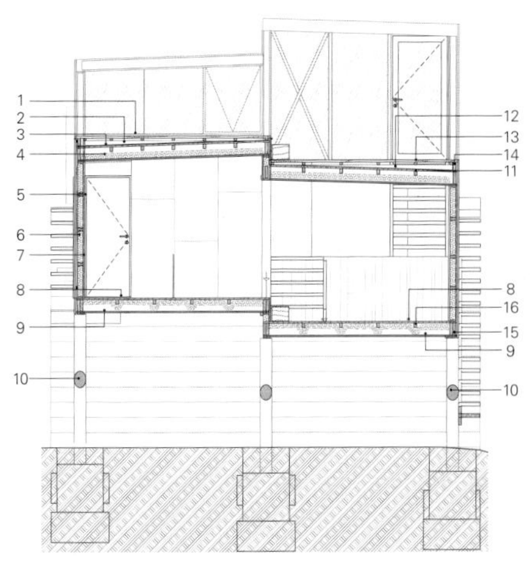

B-B' 剖面图——公共空间
section B-B'_common space

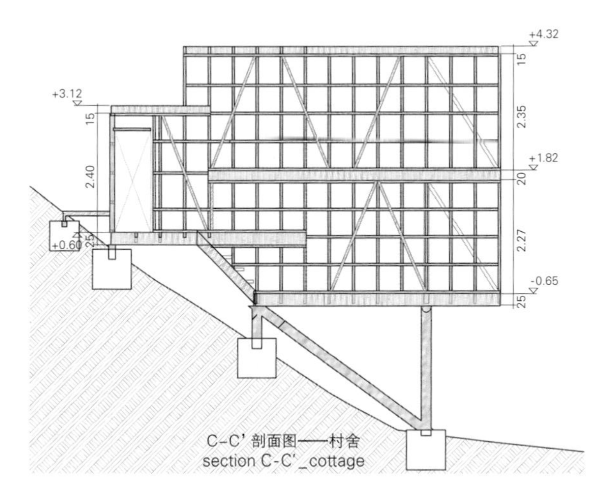

C-C' 剖面图——村舍
section C-C'_cottage

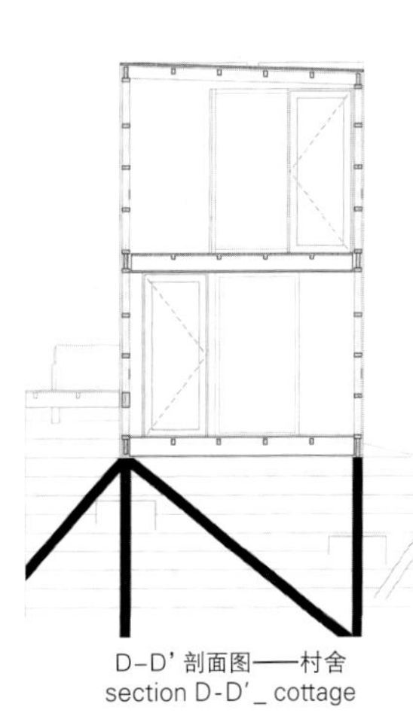

D-D' 剖面图——村舍
section D-D'_cottage

1. roof floor: pine board 1.5x4", 20mm separation, carbonileo painting
2. asphalt membrane(3% slope), 18mm plywood
3. floor and roof insulation 80mm
4. ceiling interior coating: pine board 1.5xvariable coating, vapour barrier
5. exterior wall coating pine board 1x4" with Junquillo pine wood 1x1", carbonileo painting
6. wall insulation polystyrene 40mm
7. wall interior coating pine board 3/4"x variable width
8. pine board dovetailing floor 1x4" vitrified
9. 18mm plywood
10. impregnated wood buxom 8"diameter
11. pine link 2x3"
12. plywood 15mm roof
13. pine strio 2x2" for terrace boards
14. impregnated wood cornice 2x6"
15. pine wood beam 4x10"
16. pine wood strip 2x3"

Buildings for ki

从教育角度看幼儿园建筑

社会的不断发展和变化，使得教育成为一个日新月异的领域。当代教育方法倾向于囊括新型通讯和信息技术。其中包括互联网、多媒体资源或需要全新的教室硬件条件的新型技术设备。现有的旧建筑就应该适应这些新活动，而那些为了满足这些新要求而设计的新建筑，则更应灵活地响应当前我们所面对的技术应用的快速变化。

而且，现如今城市中心的生活引导孩子和学生们有更多校园生活的时间。如此一来，就产生了提供新型教育和教育外空间的需求，并让人们重新从全局考虑学校，尤其是教室的作用。

除去那些横向条件，不同的教育方法和文化也影响着建筑设计：一个采用蒙台梭利教学方法的学校在结构上明显不同于基于音乐或表演艺术教学方法或与宗教信仰相关的学校。空间的设计目的如下：用于个人或集体工作；用于动手实践、示范或讲解体验；用于实验、理论或运动活动等——只有实现精确的基础设施系统，并选择合适的外形、颜色、材料、照明设备等元素，才能回应这些复杂的用途要求。

Teaching is an ever-changing field, as the society is in constant development and mutation. Contemporary teaching methods tend to incorporate new communication and information technologies. These can include the internet, multimedia resources or new technological devices that demand new conditions from the classroom. Existing old buildings should adapt to the new activities, and the newly constructed ones, already planned for these new requirements, should be flexible to respond to the fast changes of technology's uses we face in the modern days.
Also, life in the urban centers today is leading children and students to spend more time in school. This comes along with the need of providing new educational and non-educational spaces, and to rethink the role of the school, in the large scale, and of the classroom, in particular.
Besides those transversal conditions, the different teaching methods and cultures also reflect on the building design: a Montessori method school has structural differences from a music-based or performing arts-based teaching method or a religion-related school. Spaces are conceived for individual or collective work; for hands-on, demonstrative or expositive experience; for laboratorial, theoretical or kinetic activities, etc. — a complex set of requirements that architecture must respond to by implementing an accurate infrastructural system and an appropriate choice of forms, colors, materials, lighting, among others.

从教育角度看幼儿园建筑_Buildings for Kids as Educators/Paula Melâneo

s as Educators

德国心理学家罗特劳特·瓦尔登在她撰写的《未来学校：从建筑心理学看设计方案》（2009年）一书中说道："学校建筑是除教育者和同学之外的'第三位老师'。"

这也正是长久以来解释以下情况的强有力的理由：教育心理学研究一直伴以建筑在教学环境空间构型中的作用的研究。

根据罗特劳特·瓦尔登的观点，我们可以根据若干要求分析不同的学习环境，例如：颜色方案、形式设计、照明、供暖、制冷和通风；隔声和噪声；家具和设备；密度和拥挤度；参与度与使用者设计；生态学影响；学校的组织因素或课外使用。

空间条件是健康学习和全面认知发展的重要因素，而建筑本身应该像教师和其他教育者那样以身作则，在这里良好的实践代表了可能遵循的联系。这种情况应始于建筑项目之前，通过为具有包容性和参与度的设计创造条件，使与学校关系最密切的社会和利害关系人参与到对建筑物的期望和使用需求的讨论当中，如此一来，新建筑就会提供一个理想的解决方案，并随着时间的流逝而更加经久耐用。

智能设计、安全的循环通道设施和有效的空间组织对于促进普通使用者和残障人员之间积极的社会关系是不可或缺的。而考虑室内和室外空间，致力于娱乐和学习活动，以及在分离区域和办公室为教师提供私人空间同样也很重要。

大型空间并不一定意味着高品质，因为它们可能体现了一种越权行为，并与孩子/学生们拉开距离，并可能导致设计无个性特色，而一个舒适、安全的空间才是最为适合的设计。

建筑外形、材料和颜色与周边环境完美融合，平衡环境中的影响，帮助社区居民接受这个新生事物，也避免使用者在短期之内对整个基础设施产生审美疲劳。

设计师应该特别注意使用易于清洁和无毒的材料，特别是室内设计，也要注意利用模块化和易于挪动的家具，以加强空间的灵活性。另一个积极的过程在于为用户提供对环境压力因素的控制权限，使用户能够自己调节照明、通风、供暖、雨水防护以及隔声效果等不同系统。

学校建筑的设计应从可持续角度出发，考虑其潜在的未来改造或

"The 'third teacher' besides the educator and the follow student is the school building.", says Rotraut Walden, German psychologist, in her book *"Schools for the Future: Design Proposals from Architectural Psychology"*(2009).

This is a strong reason why, over time, the educational psychology studies have been accompanied by the study of the architecture's role in the spatial configuration of pedagogical environments.

According to Rotraut Walden's ideas, we can analyze the different learning environments following several requirements, such as: color scheme; form design; lighting; heating, cooling and ventilation; acoustics and noise; furniture and equipment; density and crowding; participation and user design; ecological aspects; organizational aspects or after-hours use of schools.

The space qualification is essential for a healthy learning and a thorough cognitive development, and buildings, such as teachers and other educators, act themselves as examples, where the good practices represent possible links to follow. Such fact can start just before the building's project, by creating the conditions for an inclusive and participatory design, engaging the school's closest society and stakeholders in the discussion about wishes and needs, so the new structure will be able to offer the ideal solutions and last longer in its use over time.

Intelligent design, safe circulation facilities and effective spatial organization is fundamental to promote positive social relations between users and accessibility for people with disabilities. It is also important to consider indoor and outdoor spaces, dedicated to playful activities and to learning, as well as providing private spaces for teachers, in separated areas and offices.

Larger spaces don't always mean quality because they may reflect an excess of authority and distance from the children/students, and can lead to anonymity - a cozy, safe and comfortable space is the most desirable.

Shape, materials and color that are well integrated in the immediate environment, balancing the impact within the surroundings, facilitate the acceptance of the newness by the community, avoiding that the users get tired of the whole infrastructure in the short-term.

A careful attention should be given in the use of easy clean-

Drawing©CO-AP

发展（例如区域扩展、课外的多用途使用、用途的改变等等），其灵活性可能会在较长的预期使用期限内受到局限。可以从生态的角度考量建筑项目以及施工，引入责任用户行为，使用耐用、易维护的环保型或再生材料，并通过评估合理的消耗方法和高效运营——智能通风、照明和供暖系统。其用途可预见资源的重复利用，如重复利用废水和/或雨水，或包含可收集和循环再利用日常废料的存储空间。

本章介绍的项目均为现代化的设计方式，与构思一个理想的学习空间的复杂性相互呼应。它们一般都采用了素淡的色彩设计方案，使用自然色调而非对比强烈的色彩，在内部和外部空间中均采用丰富的纹理和材质，采光充足，并按照儿童适宜的尺寸设计，实现安全而舒适的结构，为新型学习体验创造了多变的激励空间。

扩建并改造现有建筑

蒙特西奈幼儿园位于墨西哥城，是原有学校大楼的扩建建筑，占地3755m²，由LBR+A事务所设计完成。这栋全新的大楼包括一家幼儿园，用作现有小学、中学和高中大楼的附属建筑。其特殊性在于由于空间有限，该建筑位于校区现有的即停即离区和停车场的顶部。在同一水平上还规划设计了一座礼堂和一座体育馆。

在该水平面之上，是几处两层高的包含教室单元的楼群。不同的颜色使得它们彼此独立并且更好区分，它们看起来就好像是一些大型的玩具。包含自然元素（植物、草地、树木等）的邻近天井、露台和庭院是能够支持课间活动的户外空间。整体规模设计符合儿童使用舒适度，家具的设计比例合理，并且正如建筑师所解释的，"这些空间让人产生了一种强烈的归属感"，所以孩子们在空间里会产生舒适和安全的感觉。

考虑到节约能耗，新建筑物采用朝南设计，允许阳光直射进入教室当中，使得室内和室外区域达到舒适的温度。施工进程充满灵活性，其特点是在停车场区域的上方架起一个金属框架结构。使用预制嵌板铺盖在教室的外墙面，这是保持适宜室内温度的一个快速且简单的施工方案。同样，从生态效益角度来看，该设计包括雨水、废水的收集和净化结构，所以可以对废水进行再利用，这对建筑本身也是一个具有教育意义的设计提案。

ing and non-toxic materials, especially in the interiors, and of modular and easy-movable furniture, potentiating the spaces' flexibility. Another positive procedure is providing user control of environmental stress factors, giving the ability to personally regulate the different systems of lighting, ventilation, heat, rain protection, acoustics, etc.

The school building should be designed with a sustainable perspective, thinking about its possible future transformation or evolution (for example expanding areas; multiple uses besides class time; change of uses; etc), where its flexibility can be traduced in a larger lifespan. The building's project, and construction can be thought from an ecological point of view, introducing a responsible user behavior, by using durable, low-maintenance, environmentally friendly or recycled materials; by estimating a reasonable consumption of means and operate energy-efficiently – intelligent ventilation, lighting and heating systems. Its use can foresee the reutilization of resources, like reusing grey water and/or rainwater, or include dedicated spaces for collecting and recycling daily wasted materials.

The projects presented in this chapter are contemporary approaches responding to the complexity of conceiving an ideal learning space. In common they have a sober chromatic approach, regarding the use of natural tone palettes rather than using strong bright colors, the richness of textures and materials – both in inside and outside spaces, the generous entry of sunlight and are carefully projected as children-friendly scales and as safe and comfortable structures, with varied stimulating spaces for new learning experiences.

Extending and Transforming Existing Buildings

The Kínder Monte Sinai is a 3,755 square meters addition to an existing school complex in Mexico City, designed by LBR+A. This is a totally new building housing a preschool, functioning as a complement to the existing elementary, middle and high school buildings. Its particularity is to be located in the top of the existing drop off and parking area of the campus, due to the lack of space. In the same level an auditorium and a gym were planned. Over this level, several blocks house the classrooms' units in two levels. Different colors make them more detached and identifiable, as if they were large toys. Contiguous patios, terraces and courtyards with natural elements (plants, grass, trees, etc) are the outdoor spaces that can support the classes' activities. The global scale is children-friendly, the furniture is designed proportionally, and "the spaces generate a greater sense of belonging" – as explained by the architects, so the kids feel comfortable and safe in it. Considering lower energy consumption, the new structure is oriented facing south, allowing direct sunlight inside the classrooms and comfortable temperatures in the interior and

在一个更小的干预项目——占地588m²的坎普尔顿幼教中心的设计中，澳大利亚建筑事务所CO-AP进行了一次有趣的操作：将先前一处工业仓库建筑改造成一处可容纳80名儿童的幼教中心。

建筑师特别留心材料的选用，他们更为重视材质的丰富性，而非采用鲜艳的颜色，旨在激发孩子们的多种感受，并同时营造一种设计者所说的"熟悉的居家感受"。

在现场，建筑师事先查看了快速经济型的施工建设。施工采用模块化系统，并在未完工的定向刨花板上设计了很深的压印。

在现有的屋顶之下，仓库的开放空间被几个木质的"小房子"填满，每个小房子都带有玻璃窗，建筑师将它们构思设计为室内空间。这些房子中间是游戏空间，在这里，阳光透过屋顶的透明玻璃能够照射进来，营造一种置身户外的感觉。人们可以从仓库的一些断瓦残垣中感知到它的前身，强调建筑前身的重要性。

建筑师团队选择简单的形状来培养孩子们的想象力，并且与景观设计师一起营造了一个游戏体验区——在这里他们打算"提高风险评估"，使用不常用的材料，如砂岩块作为休息凳。

重新开始设计

比利时的Lens° Ass建筑师事务所在圣吉尔的市区结构中设计了圣吉尔小学。新建的六层建筑朝向Engeland大街，占据了邻近区域两栋建筑用地的空间，采用了一种现代建筑语言，不刻意与周围环境融合，但尊重其标准，并考虑到了在和谐对话中的"形式语汇"。

该设计是另一个鲜艳明亮的色彩（曾经被认为更适于激励孩子们学习）被更加中性的色彩所代替的案例，而这些颜色主要由所用的木材、金属、砌砖、混凝土和玻璃材料引入。这种方式避免了视觉混乱的情况，也为丰富多彩的教育资源敞开空间，并放飞如纯洁白纸一般的孩子们的想象力。

尽管该项目采用了多层建筑的设计形式（并不是理想的学校项目设计方案），但它在空间组织上十分简单，并且建筑师力图为孩子们提供更大的空间，"建筑设计力图实现紧凑的建筑风格，节约空间，以尽量创造更多的开放空间"。同时，在通达性方面，建筑师将成年人的办公室设置在高层区域，这样孩子们就可以在低楼层活动了。

建筑的后方是一处由高墙围合起来的大型空间，这里整合了一处

exterior areas. The construction process is thought to be flexible, featuring a metallic frame structure over the parking area. Precast panels were used to cover the classrooms' outer walls, a fast and easy constructive solution that allows maintaining an inner mild temperature. These panels were easy to perforate with circular windows, guaranteeing the natural cross ventilation. Also, from the ecological point of view, the design includes collecting and a treatment plant for the rainwater and sewage, so the water can be reused – an educational proposal from the building itself.

In a smaller intervention, the 588 square meters Camperdown Childcare Center, the Australian practice CO-AP performed an interesting exercise: transforming a former industrial warehouse building into a 80-place childcare center.

The materials were chosen with special care, giving more importance to the richness of textures than to bright colors, with the intention of stimulating children's senses and, at the same time, creating a "familiar residential feel", as the authors say.

A fast and economic construction on site was previewed, with modular systems, where the unfinished OSB(oriented strand board) panels display a strong imprint.

Under the existing roof, the warehouse's open-space is filled with several wooden "houses", with glazed windows, conceived for the indoors spaces. In between are the playing areas, where the large transparencies of the roof let the sunshine come in, creating the feeling of being in outside spaces. Some of the remnants of the warehouse are perceived in the structure, giving importance to the precedent life of the building.

The architect's team opted for simple shapes to foster children's imagination and, together with the landscape architect, created an experiential area for playing – here they intended to "promote risk evaluation", using uncommon materials, such as sandstone blocks as benches.

Designed from Scratch

The Belgian practice Lens° Ass Architecten created the Sint-Gillis Primary School within this municipality's urban fabric. Facing Engeland Street, the new six-floor building occupies the space of two building lots in the neighborhood, with a contemporary architectonic language that does not pretend to merge with the surroundings, but respects its metrics and "formal vocabulary" in a harmonic dialogue.

This is another example where the strong bright colors, once considered more appropriate to stimulate children, give place to more neutral colors, mainly introduced by the materials used – wood, metal, brick, concrete and glass. In this way the visual mess is avoided – opening space for the educational materials to be colorful and for the children's imagination, as a white paper to draw on.

Despite the program being articulated through multiple-

Drawing©Lens° Ass Architecten

露天剧场和一栋老旧荒废的建筑——使人们想起该场地先前的用途。作为一个有保护的操场，该区域的旧建筑物形成了一个带顶的玩耍空间，好似一个空旷的让人充满遐想的大房子。

在UID建筑师事务所的这个案例中，木材作为设计的主要元素，以其自然的色调主导了整体构造，它就是位于日本广岛市、占地120m²的"花生"幼儿园。本项目的名称意指花生的果实，得名于建筑的形状——产生交汇点的两个圆圈。设计形成了一个圆形的没有转角的开放空间，给人一种无拘无束的空间感，然而却能包围并且拥抱孩子们。这种形状和小规模使得学校的教职人员可以统览全局。建筑师实行空间功能分区设计，这样孩子们和员工们就能以相同的方式分享同一个开放空间了。

建筑师意在营造一种置身于森林中的氛围，在那里室内与周围的花园柔和地相融，建筑透明的外表面形成了一种延续性。建筑内部的高差较为柔和，好似自然地势的斜坡，使空间变得更加有趣，在这种设计中工作空间的平面布局可以被转变成游戏室的布局。立面上水平木材表面的设置使得阳光好似穿过树叶般进入建筑内部。在这里孩子们会感觉很安全，仿佛被建筑拥抱在怀中。

位于里布尼察的这座占地4500m²的幼儿园是为400名儿童设计的学习基础设施，它是斯洛文尼亚最大的幼儿园。外形和规模上的特点使它成为一栋小号的城区建筑物，建筑师受里布尼察市中心的城市结构和建筑的启发，设计了一些具有不同高度的独立的小房子。

针对该项目，Arhi-Tura d.o.o建筑事务所设计了一种U形的空间联系，在设计中孩子们的教室"环抱着位于中央的操场，仿佛拥抱孩子的双手，并形成了一个安全温暖的空间和环境"。在城区规划理念中，这些操场相当于公共广场，孩子们可以在与周围环境和景观直接产生联系的环境下适应社会生活。

在这所幼儿园中采用了大批不同颜色的材质——石料、草地、玻璃、不同式样的木料等，丰富了孩子们的空间体验。

除了设计得一板一眼的行政机构的建筑体量之外，孩子们教室的设计并不是直线型的，而是采用Z字形设计，以不同角度处于最佳的阳光、景观和声音条件下。

白水托儿所是由日本山崎健太郎设计工作室设计的一栋极有特色

story (which is not an ideal solution for a school program), the space organization is simple and the architects seek to provide larger spaces to the children "the architecture is meant to be compact and space-saving in order to create as much open space as possible". Also, regarding accessibility, adults have their offices on the top floor, so the children can occupy the lower floors, according to their ages.

The back of the building is a large space surrounded by high walls, that integrates a new open-air theater and an old ruined building – evoking the former use of the site. This area functions as a protected playground where the old structure acts as a covered space for playing, like an empty large house for the imagination.

An example where the wood is the main element, with its natural tone dominating the construction, is the Peanuts, a 120-square-meter nursery school in Hiroshima, projected by UID Architects. The project's name, alluding to the fruit, comes from its plan shape, the intersection of two circles. It creates a round open-space with no corners, bringing a sensation of a non-confined space, yet surrounding and embracing the children. This shape and the small scale allow educators and the school staff to have a general vision of the whole. The space functions are zoned, so the children and the staff share the same open space, in an equal mode.

The architects intended to create the atmosphere of a forest, where the interior blends gently with the surrounding garden, in continuity, throughout the transparency of the exterior surface. The level differences inside are soft, like slopes of natural topography and turn the space more playful, as where a work plan is transformed into the ground of the playroom. The set of horizontal timber surfaces in the facade make the sunlight enter as it does between the trees' leaves. Here children can feel safe, involved and caressed by the building.

The 4,500-square-meter building for the Kindergarten in Ribnica is a learning infrastructure projected for 400 children and the largest in Slovenia. Its characteristics of shape and scale make it a city-like structure in a small size, inspired by Ribnica's central urban fabric and buildings, with its small individual houses with different heights.

For this project, Arhi-Tura d.o.o created a spatial relation in a U shape, where the children's rooms "were designed to embrace the playground in the middle, imitating hands embracing the child and forming a safe, warm place and environment." Within the city-like logic, these playgrounds act like the public squares, where children can socialize in direct relation with the surrounding environment and landscape.

In this kindergarten, a large palette of textures is offered – stone, lawns, glass, different types of wood, etc – enriching children's

的建筑。沿着场地的斜坡，建筑师在这片530m²的建筑场地上设计了一组大型台阶，并将每一阶都设计为不同的游戏区域。

秉承为60名儿童设计一所"大房子"的设计理念，其主要空间的特点在于开放性，人们置身其中能够有一个全局的视角，并且孩子们会感觉更加安全。

建筑师将周边山脉和森林的氛围引入托儿所。密集的木结构好像树木一样，其中更设计了一些小的"木盒子"用来当作卫生间的封闭空间，如同森林中的小房子。大型推拉窗形成了室内与室外视野的连接，也形成了空间的对流通风，同时也使得充足的阳光进入室内，提高室内的温度。这家托儿所还设计有露天游戏空间：北面有一个露天游戏露台，南面有一个木质的露天平台。在建筑的南立面，来自于洒水装置的水或雨水由屋顶落下，在地面的水池中形成一个小瀑布，为孩子们的玩耍带来了乐趣。而当这些水在风中蒸发时，也起到了使空间更凉爽的作用。

Rh+建筑师事务所设计了位于巴黎的一家日托中心，该日托中心是一个占地1800m²的儿童基础设施。该项目由一组建筑体量组成，以其"居家规模"彻底地融入当地工业年代所形成的城市结构（一个典型的法国低密度住宅小区）当中，而其风格与巴黎宏伟的林荫大街风格截然不同。正如Rh+事务所的建筑师所说的，其"现代简约的建筑式样"是由中性材料建造而成的，而白色已然成为其外部的主色调。采用大胆纹理的连续墙面，建筑远离街道，同时通过一天中光线的变化来形成一种动态的视觉游戏。建筑看起来好像与外部隔绝，而在内部，空间却又通过玻璃窗户相互联系。这些空间特点为孩子们带来了一种安全感，也让托儿所的教职员工能更方便地控制局面。自然光线进入内部空间；不同的建筑体量产生了露天区域；不同材料和不同颜色的应用定义了空间的用途，而这些都是孩子们平日里在托儿所生活玩耍时发挥想象力的基础。

spatial experiences.

Apart from the administration volume, with a sober geometry, the children's rooms are not linear, but zigzag-oriented, being differently exposed to optimal sunlight, views and sound conditions.

The Hakusui Nursery is an uncommon building designed by the Japanese Yamazaki Kentaro Design Workshop. This 530-square-meter surface is designed as a set of large steps that accompany the site's slope, where each step creates a differentiated area for play.

Following the idea of being a "large house" for 60 children, the main space has open characteristics, enabling a global overview and making the children feel safer.

The architects brought into the nursery the atmosphere of the surrounding mountain and forest. A dense wooden structure resembles the trees, where in between are constructed small wooden boxes for the toilets' closed spaces, just like small huts in the forest. Large sliding windows make the visual connection with the exterior and the cross ventilation of the space, and, at the same time, allowing the sunlight to generously enter and heat up the interior. The nursery also has open-air playing spaces: In the north there is an open-air play terrace, and in the south a wooden deck. Here, in the south facade, water falls from the roof, coming from sprinklers or rainwater, and creates a small waterfall with a water pond on the floor, which acts as playful artifice for the children's joy. This is also a device to cool the space, with the water being evaporated by the breeze.

The office Rh+ Architecture is the author of the Day Care Center in Paris, a 1,800-square-meter infrastructure for children. This project consists of a set of volumes, with a "domestic scale", completely integrated in the local urban-fabric – a typical low-density dwelling French neighborhood, from the industrial era, with a remarkable difference from the large Parisian grands boulevards. Its "contemporary and simple architecture" – as say Rh+ studio – is constructed in neutral materials, with white as its exterior's main color. A continuous wall with a bold texture makes the separation from the street creating, at the same time, a dynamic visual play, by the light's changes along the day. The building seems closed from the exterior, contrasting with the interior, where the spaces communicate between themselves by glazed windows. These spatial characteristics generate a sense of security for the children and represent more control for the nursery staff. The natural light entering the interior spaces; the open-air areas created by the different volumes of the building; the variety of materials applied and the different colors defining the use of the spaces, are the basis for the children to imagine new narratives during their daily usage of the nursery. Paula Melâneo

蒙特西奈幼儿园

LBR + A

蒙特西奈幼儿园尺寸设计合理，环境舒适，这样的教育环境可以激发孩子们的创造力。这里增强了孩子们的归属感，为孩子们创造了一个安全舒适的坏境。新建幼儿园是在现有的小学、初中、高中学校旁边扩建的。由于校园空间有限，该项目位于学生上下车区域和停车场的顶部。它共分为三层：停车场层、二层和三层。

停车场层除了停车区域还专门为幼儿园加建了礼堂和体育馆。二层建有操场、主庭院、图书馆和二层的教室，这一层覆盖在停车场上方，视野开阔，安全性高。

教室垂直于主轴线设置，暗指积木堆砌的概念。首层的教室之间有许多小庭院，它们是为各个教室配备的，这里种满了各种植物，旁边还有通往三层的楼梯。由于三层的教室与二层教室布局一样，所以每间教室的二层与三层之间都有平台。

教室10m长、5.2m宽，高度是2.1m，为孩子们创造了一个舒适的空间。每个教室里都配有一间小厨房、储藏室和浴室。家具是根据儿童的身高而定做的。幼儿园教室共有18间，其中12间采用了预制面板覆盖，6间采用了彩色层压玻璃。

预制法是教室外墙建造的最佳选择。幼儿园朝南，白天日光充足，即便是在没有直射太阳光的时候，预制面板也能保持室内的温度稳定。由于施工场地和工作时间受到限制，采用预制法将有利于施工的进行。预制法也适用于滴灌设计、高度不同的圆形窗户设计以及优质的白漆粉刷。

停车场层的长跨度是通过空腹桁架结构来实现的。其矩形框架使得该建筑的设计更具有灵活性。一些垂直于桁架布置的教室都为悬挑结构。

该项目运用可持续设计是非常重要的。这所幼儿园朝南，孩子们在学校时拥有充足阳光，同时也减少了建筑的能源消耗，无论是教室的内部还是外部都能维持舒适的温度。预制面板上的圆形窗户能达到自然通风的目的，而建筑收集到的雨水和污水都能够循环利用。

Kinder Monte Sinai

The Kinder Monte Sinai encourages children's creativity through educational spaces designed to their size and comfort. The spaces generate a greater sense of belonging and create a comfortable and safe environment for the children. The new preschool building is an expansion of the existing elementary, middle school, and high school. Because of the limited campus space, the project is located on top of the student drop off and parking area. It is distributed on three levels; the parking level, the first floor, and the second floor.
In the parking level, besides the drop off and parking area, an auditorium and gym were added for preschool only. In the first level, there is a playground, main courtyard, library, and first floor of classrooms. This level covers the parking area and is 36 feet above, offering many benefits such as safety and good views for the classrooms.
The classrooms are placed perpendicular along the main axis, alluding to the concept of stacked toy blocks. In the first floor, between the classrooms, are small patios designated for each classroom. Behind them are courtyards with plants and trees and stairs that lead you to the second level. Since the classrooms on the second floor are overlapped, there are terraces for each one.

The classroom units are 33 by 17 feet and 6 feet 10 inches in height designed to create a comfortable space for the children. Inside the classrooms are a small kitchen, storage room, and bathroom. Furniture is designed at the children height. There are 18 classrooms in total, 12 covered in precast panels, and 6 in colored laminated glass.
The precast, being the best solution, was used to cover the outer walls of the classrooms. The building's orientation is south, receiving sunlight during all hours of the day therefore the precast maintained a steady indoor temperature without direct sunlight. Because of the limited construction site and the restricted working hours, the precast facilitated construction. The precast allowed the design of drippers, round windows placed at different heights, and a white washed quality finish.

The long span to cover the parking on the ground level was achieved with the vierendeel truss. Its rectangular unobstructed frame permits flexibility for the architectural design. Some of the classrooms, aligned perpendicular to the truss, are in cantilever.
It was very important to attain a sustainable design for the project. The preschool, positioned facing south, receives direct sunlight during school hours, reducing energy consumption and maintaining comfortable temperatures inside and outside the classrooms. The position of the round windows on the precast panels allow for natural cross ventilation. Rainwater and sewage is collected and treated in the water treatment plant to be reused.

项目名称：Kinder Monte Sinai
地点：Av. Loma de la Palma 133, Colonia Vista Hermosa, Delegación Cuajimalpa de Morelos, México, DF
建筑师：LBR+A
合作方：architect_Benjamin Romano / 设计团队：_Julieta Boy, Aby Helfon, José Luis Martin, Mariana Mercado
施工方：ANHOS
结构设计：VAMISA+DITEC / 电气工程：Uribe Ingenieros
混凝土供应商：CEMEX / 立面：Grupo HEG / 金属结构：Corey
预制产品供应商：PRETECSA / 客户：Colegio Monte Sinaí
总建筑面积：3,755.70m²
施工时间：2012.9 / 竣工时间：2013.12
摄影师：©Alfonso Merchand (courtesy of the architect)

1 教室 2 露台 3 实验室
1. classroom 2. terrace 3. laboratory
三层 second floor

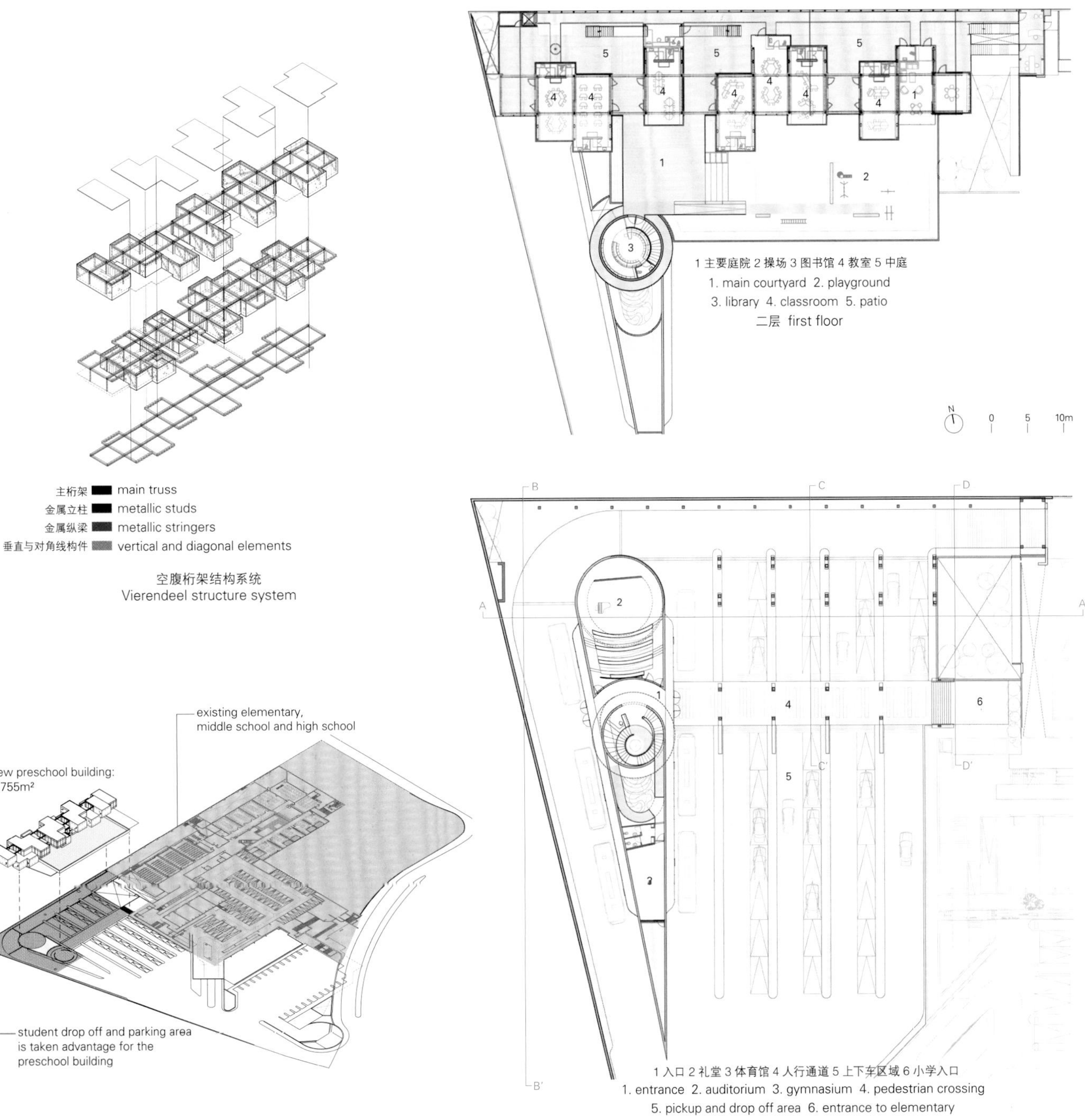

空腹桁架结构系统
Vierendeel structure system

1 主要庭院 2 操场 3 图书馆 4 教室 5 中庭
1. main courtyard 2. playground
3. library 4. classroom 5. patio
二层 first floor

1 入口 2 礼堂 3 体育馆 4 人行通道 5 上下车区域 6 小学入口
1. entrance 2. auditorium 3. gymnasium 4. pedestrian crossing
5. pickup and drop off area 6. entrance to elementary
一层 ground floor

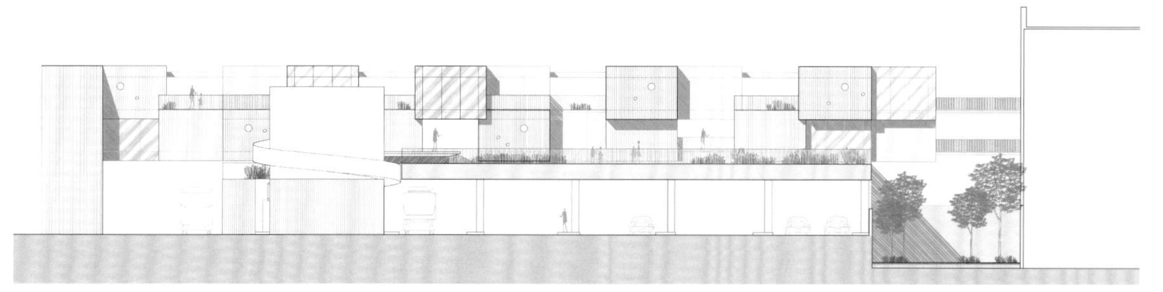

南立面 south elevation

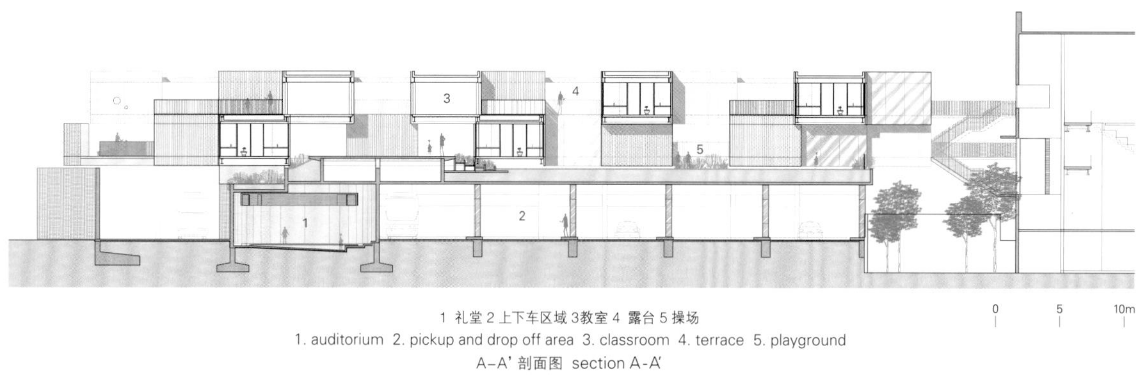

1 礼堂 2 上下车区域 3教室 4 露台 5 操场
1. auditorium 2. pickup and drop off area 3. classroom 4. terrace 5. playground
A-A' 剖面图 section A-A'

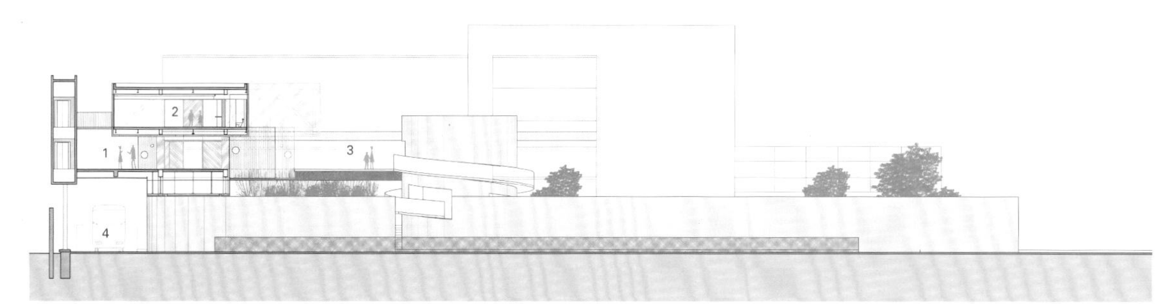

1 中庭 2 教室 3 主要庭院 4 停车场
1. patio 2. classroom 3. main courtyard 4. parking
B–B' 剖面图 section B-B'

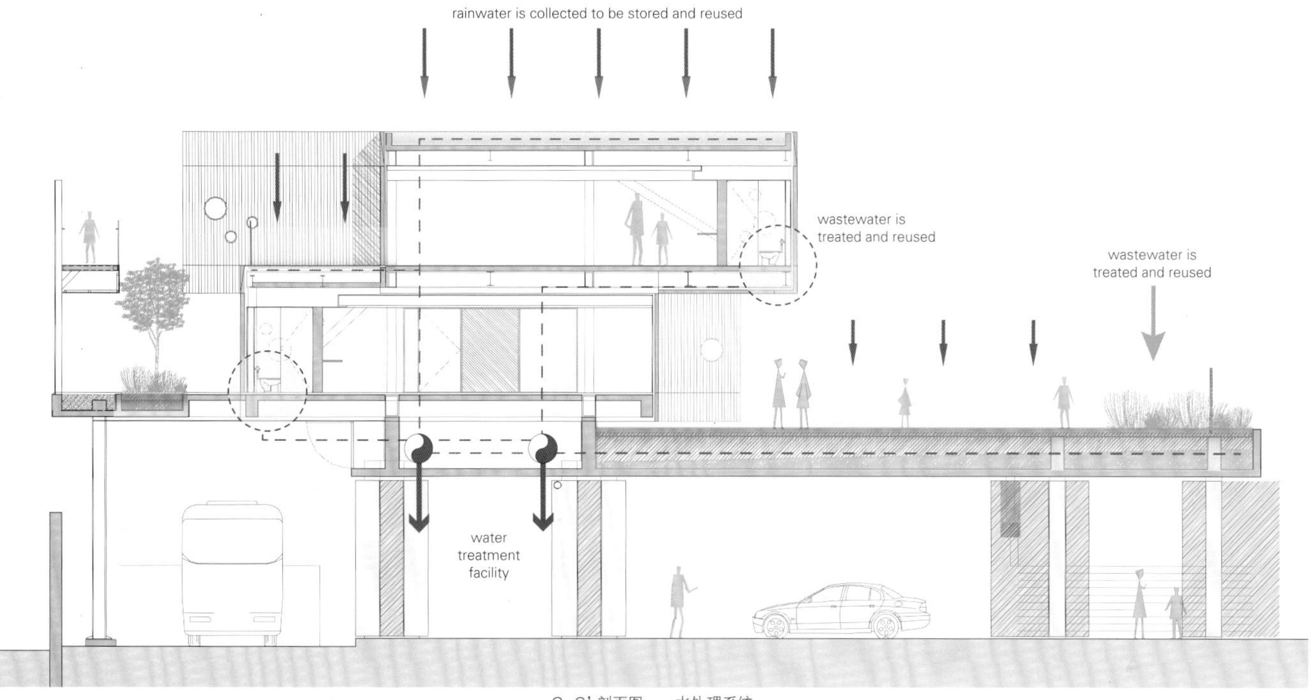

C–C' 剖面图——水处理系统
section C-C'_water treatment

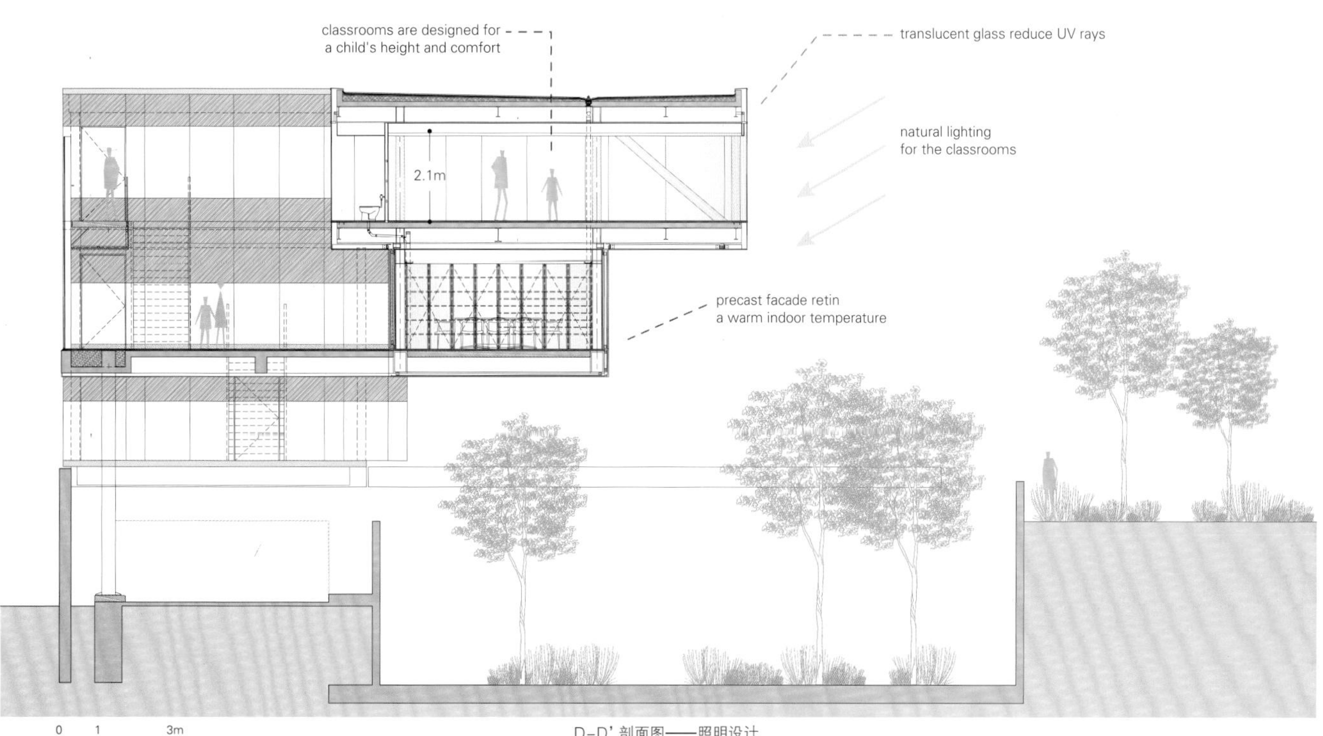

D–D' 剖面图——照明设计
section D-D'_lighting

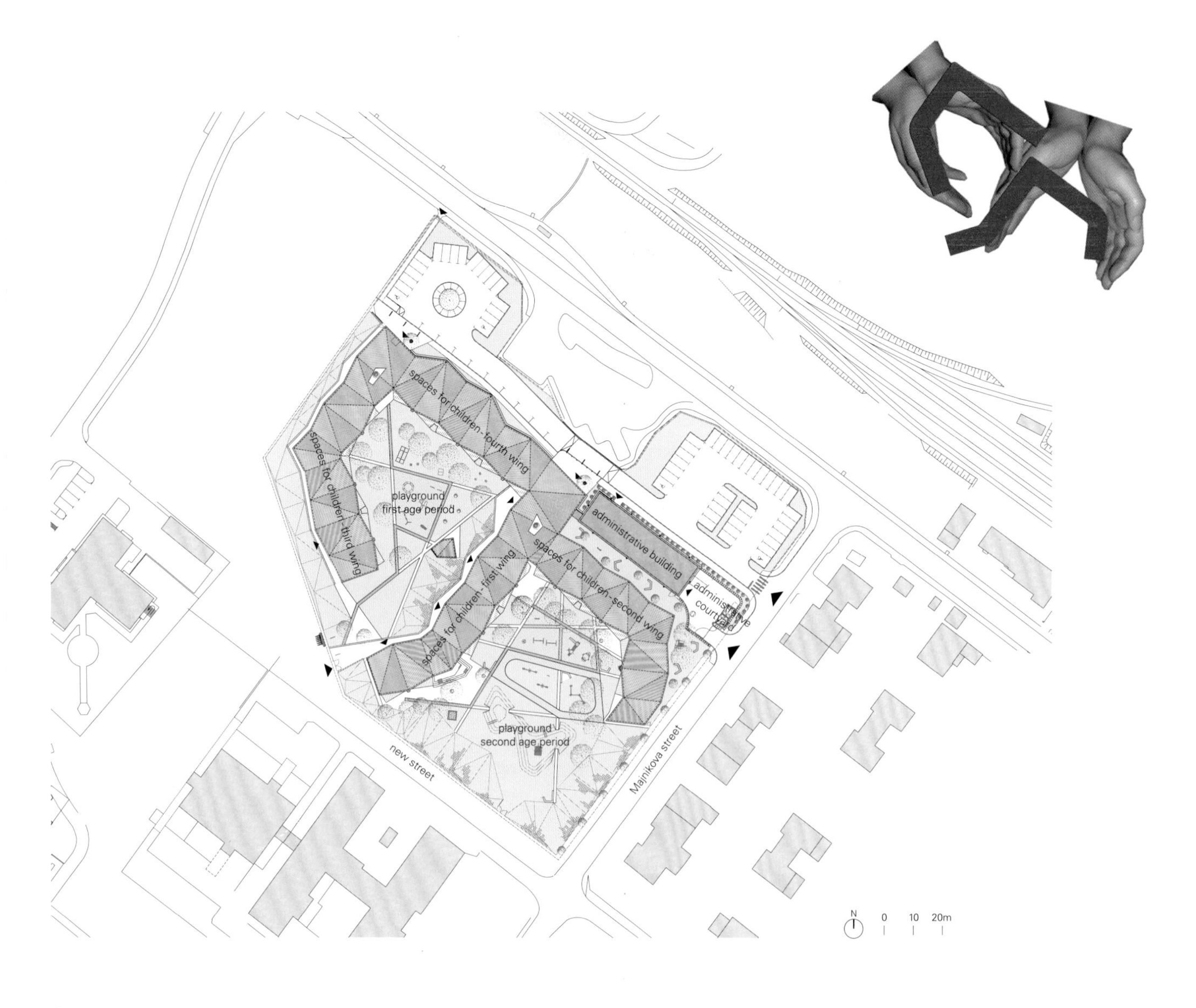

里布尼察幼儿园

Arhi-Tura d.o.o

我们想让里布尼察幼儿园成为这样一个幼儿园，既是孩子们游戏的乐园，也可让建筑师们发挥设计才能。向您介绍的这座新建的里布尼察幼儿园是斯洛文尼亚最大的幼儿园。该项目获得了建筑大赛的一等奖。幼儿园采用天然材料建造，几乎是零能耗建筑。

幼儿园的房间环绕着中间的操场，这是模仿孩子们拥抱的手势设计的，打造了一个安全、温暖的环境。为较小的孩子准备的房间封闭性较好，几乎完全将孩子们围绕在内，而另一些紧靠在一起的房间并非完全封闭，是为较大的孩子准备的，当条件允许时，他们可以外出并拥有更广阔的空间，因为设计者非常希望幼儿园的孩子们一生都拥有安全感和温暖。

幼儿园建筑的平面布局大致形成两个部分，像是两个U形也像是两个口袋的形状抑或两个海湾，每一个部分的朝向都是不同的。该建筑的线条十分曲折，尺寸适应孩子们的体型以及该地块的特点。考虑到阳光、视角及是否会受噪声侵扰等因素，分别对每个房间的朝向进行了优化改进。此外，考虑到周围现有建筑以及自然环境特征等因素，我们对建筑物的空间布局进行了深思熟虑的处理。里布尼察市中心的现有建筑网络给了我们很大启发，这里的居民住宅一个接一个沿着道路分布。这些房屋之间有着细微的区别，比如屋顶高度的不同。它们组在一起就形成了一片形状不规则的网络。

孩子们使用的楼与幼儿园行政楼形成了鲜明的对比。儿童房间的设计旨在营造一种趣味性强、有创新性且温馨的生活方式。两栋楼的设计角度不同，却彼此迎合，构成了两个海湾抑或口袋的形状围绕着中间的高品质操场。楼层平面图与立面及屋顶这些建筑元素以不同的角度彼此相交，方式不同，却交相辉映，共同构成了一件建筑作品。

走廊穿插于两侧的游戏室之间，不但可提供光源（北面的光源）还面向自然。同时，走廊的宽度可调节，也可起到游戏室的作用。此外，室内天花板与屋顶的线条相契合，从而创造出各种高度不同的房间。衣帽间和卫生间的天花板较低，但游戏室的天花板较高，这样可以延伸空间。我们在游戏室的最高处设置了一个折叠式平台，使房间更具吸引力。走廊大大拓宽，为中央区域获取了空间。中央房间的部分地面如同阶梯呈下降趋势，营造出一种剧院般的环境。长长的建筑立面将中央区域朝向公园开放。室内外的设计采用了各种各样的形状、颜色、材料、纹理与灯光。它创造了一个非常丰富的环境，为孩子提供各种各样的体验。

Kindergarten in Ribnica

Let the kindergarten become a place where an architect's play with space blends with a child's play. We are presenting the newly built kindergarten in Ribnica which is the biggest kindergarten in Slovenia. The project won the first prize in a architectural competition. It is built with natural materials and it is an almost zero-energy building.

The rooms of the kindergarten were designed to embrace the playground in the middle, imitating hands embracing the child and forming a safe, warm place and environment. Rooms, intended for children of early ages, almost fully embrace the children, while a hug of the rooms, guarding children of later ages, is less closed and points to the children's leaving for a big, wide world when the time comes. Therefore, authors earnestly hope that the feeling of safety and warmth, given to the children in the kindergarten, will accompany them throughout the life.

The floor plan of the building roughly forms a shape of two 'U' letters or two pockets or bays, each having a different orientation. The building's line is strongly zigzagged and adapted to the child's size as well as the plot's features. The orientation of certain rooms is optimized for each room separately, depending on exposure to the sun, views, noise, etc. Further, great care has been taken at placing the buildings in space, taking into account existing buildings in the surroundings as well as natural characteristics. The inspiration was found in the existing network of buildings in the center of Ribnica, where individual houses, one by one, follow the road. There are small gaps between individual houses as well as slightly different roof heights. Altogether they form a long network of irregular shapes.

Children's building has been designed to contrast with the administrative-commercial building. Rooms for children are designed in a much playful, crumbling, and attentive manner. Therefore, they follow each other, under different angles, and forming together two big bays or pockets that surround and form a quality playground environment in the middle. Architectural elements – on the floor plan as well as on the facade and roof – meet each other under different angles, in different ways and jointly creating an architectural play.

The corridors have been designed along broken playroom wings providing light (northern light) and facing the nature. At the same time, their variable width can serve as a playroom. Moreover, the indoor ceiling follows the line of the roof, thus creating rooms with different heights. Ceiling lowers in cloakrooms and toilets, whereas rises in playrooms to open the space. In the highest part of the playroom, we have placed a folding platform giving additional appeal to the room. The corridor widens considerably, making space for the central area. A part of central rooms' floor descends terrace-like and creates a theater environment. The long facade opens central area into the park. Both the interior and exterior are designed with variety of shapes, colors, materials, textures, lights. It creates a very rich environment to provide the child with various experience. Arhi-Tura d.o.o

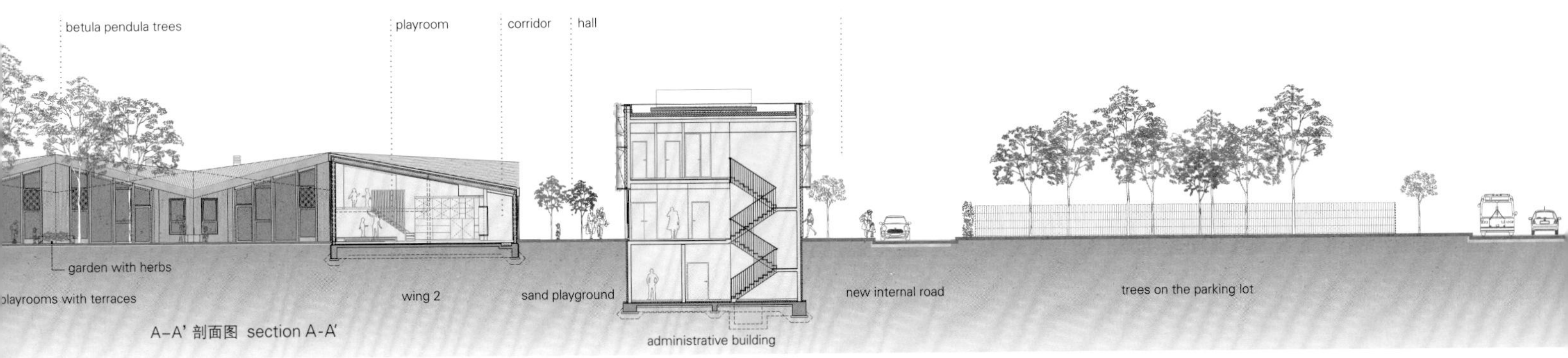

A-A' 剖面图 section A-A'

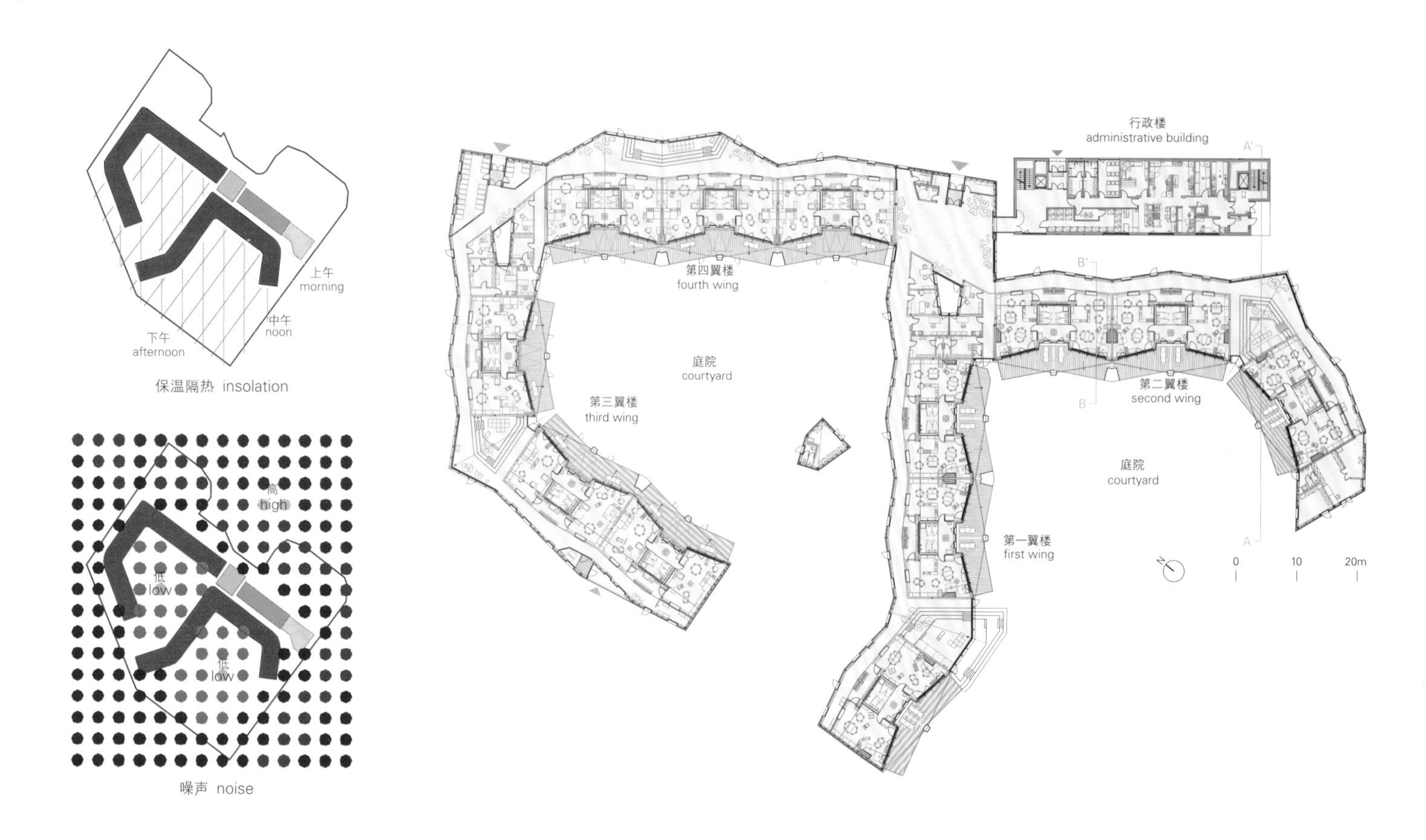
上午
morning
中午
noon
下午
afternoon
保温隔热 insolation
高
high
低
low
低
low
噪声 noise
行政楼
administrative building
第四翼楼
fourth wing
庭院
courtyard
第三翼楼
third wing
第二翼楼
second wing
庭院
courtyard
第一翼楼
first wing
0
10
20m

1. 2mm metal sheet in 600mm stripes with 25mm standing seams / 25mm wooden paneling / 40/100mm wooden battens / vapor-permeable waterproof membrane / 80mm vapor-permeable waterproof wood fibreboard / 240mm mineral wool / 100/160 mm wooden rafters, steel bearers / 50mm steel studs / 12.5mm gypsum plasterboard **2.** insulated glazing: double glazing in larch frame **3.** 12.5mm gypsum plasterboard / 20mm oriented-strand board / 75mm steel studs / 5mm sound barrier membrane / 75mm steel studs / 2x12.5mm gypsum plasterboard **4.** 30/40mm vertical larch battens, untreated / 40/60 mm horizontal wood counter battens / 60/20mm vertical wood battens / black vapor-permeable waterproof membrane / 40mm vapor-permeable waterproof wood fibreboard / 60/150 mm laminated timber studs / vapour barrier / 50mm steel studs / 2x12.5mm gypsum plasterboard **5.** 3mm polyurethane coating / 62mm micro-reinforced screed / 65mm elements of thermal heating / 40mm expanded-polystyrene insulation **5.** 20mm impact-sound insulation / waterproofing membrane: polymer-bituminous, two layered / 300mm reinforced concrete slab / 40mm lean concrete **6.** 10mm ash tree parquet / 50mm micro-reinforced screed / 65mm elements of thermal heating / 40mm expanded-polystyrene insulation / 20mm impact-sound insulation / waterproofing membrane: polymer-bituminous, two layered / 300mm reinforced concrete slab / 40mm lean concrete **7.** laminated larch timber column

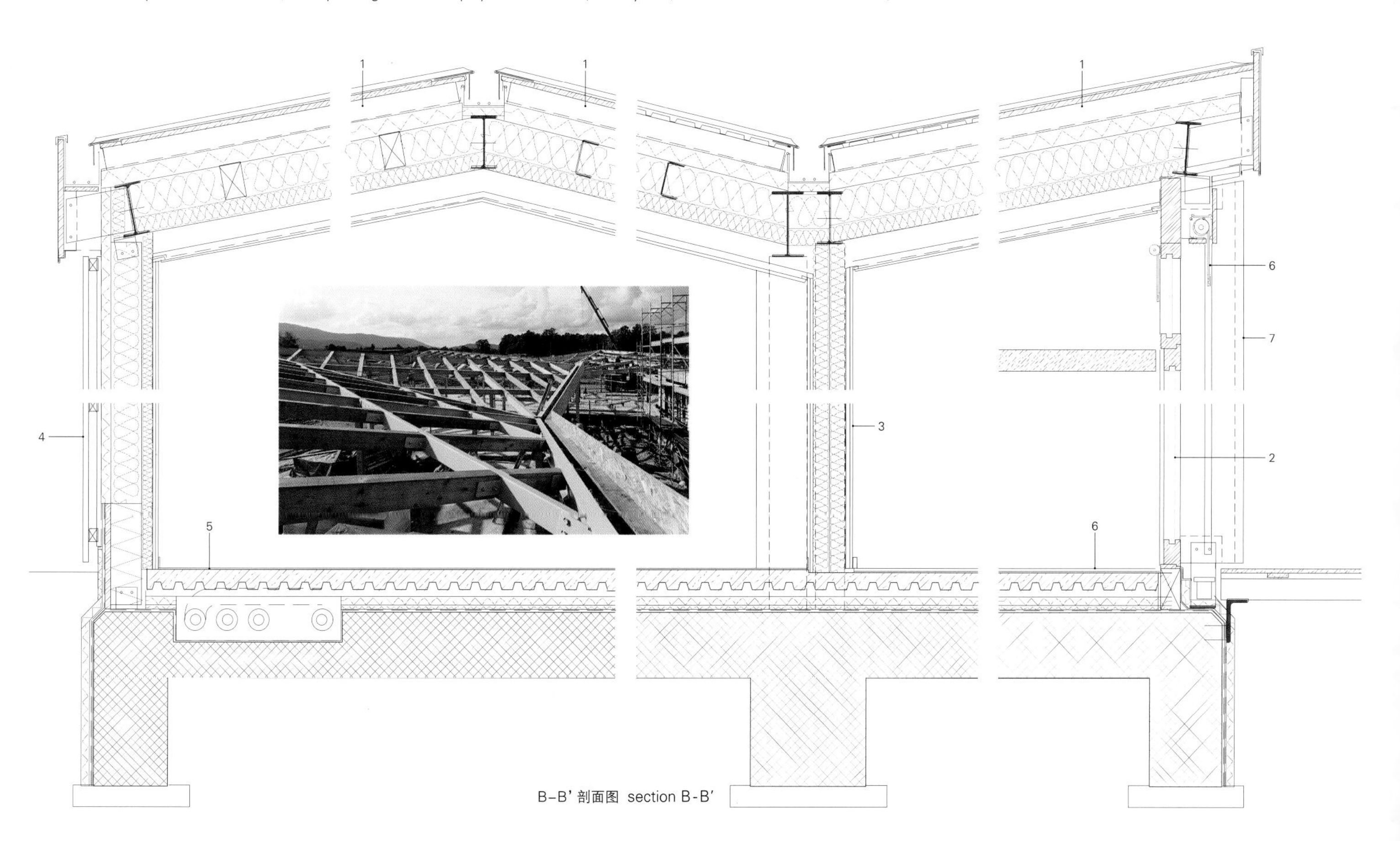

B–B' 剖面图 section B-B'

幼儿园儿童体验高差的各种场景
possibilities of experiencing the change of height for children in the kindergarten

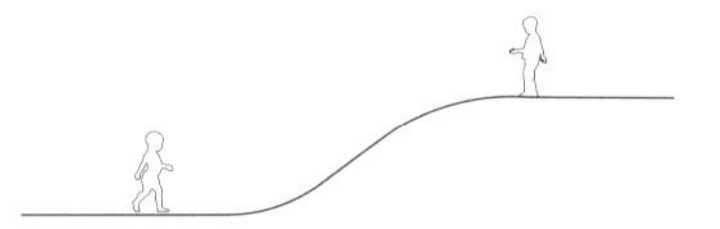

操场地面升降
rise of terrain on the playground

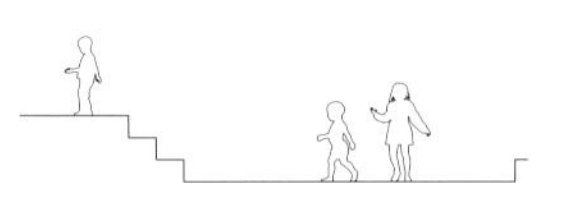

公共空间环形剧场形式的地面下降
amphitheater style floor deepening in common space

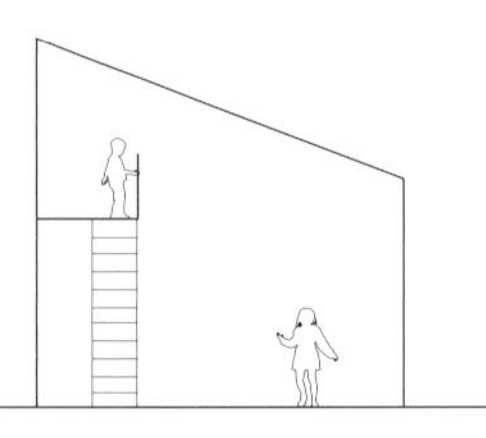

游戏室的高架平台
raised platform in the playroom

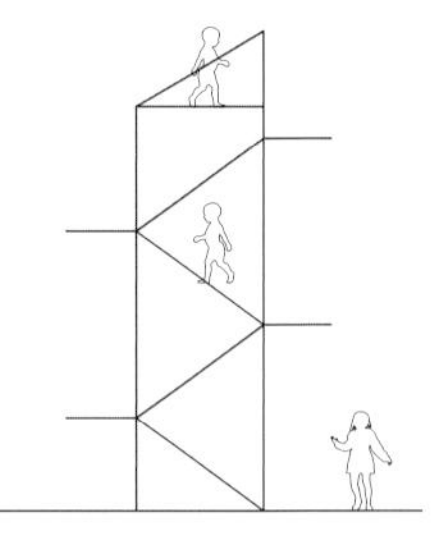

操场的游戏塔
tower on the playground

项目名称：Kindergarten in Ribnica / 地点：Majnikova 3, Ribnica, Slovenia / 建筑师：Arhi-Tura d.o.o.
项目团队：Bojan Mrežar, Renato Rajnar, Peter Rijavec / 景观顾问：Landscape d.o.o.
结构工程师：Igra d.o.o. / 机械项目：PKK Biro d.o.o. / 电气项目：PKK Biro d.o.o.
照明设计顾问：Soncesenca / 绿色建筑顾问：Mitja Košir / 施工：Riko d.o.o. / 施工造价：EUR 8m
用地面积：14,973m² / 建筑面积：4,392m² / 总楼面面积：4,946m²
设计时间：2010 / 施工时间：2012—2014 / 竣工时间：2014
摄影师：©Jorg Ceglar (courtesy of the architect)

白水托儿所

Yamazaki Kentaro Design Workshop

这家托儿所位于日本千叶县佐仓市，原计划容纳60名学生。西佑凯是当地一家社会福利公司，专门运营老年人保健设施，他们为这个项目前来与我们接洽。这个项目的首要理念来源于一个想法：幼儿园是一所大房子。场址南部是一个缓坡，被群山和森林环绕着。我们利用这种地形设计了教室，看起来像是一组大型的楼梯。

这幢"大房子"有一个独特的特征，便是3岁的孩子与5岁的孩子是在同一个房间的，在这个宽敞的大房间里，这些不同年龄的孩子们可以相互交流。另外，如果3岁的孩子睡了但5岁的孩子仍在附近玩耍该怎么办？这种不同的生活节奏就要求这所"大房子"的质量必须过关。为了与西佑凯的长期（26年）发展理念相一致，我们优先考虑最小化室内盲区的数量，这样就可以尽量保证安全措施不被破坏了。利用南北侧立面上的大型推拉窗和斜坡空间使室内空气流通，当微风在从建筑南面进入，穿过如森林般的内部立柱进而向上流动，再由北面露台而出。

南侧有个小池塘可以收集雨水与安装在南向倾斜的屋顶上的花洒喷出的水，有利于空气流动，因为吹入建筑的风将这些水气吸纳并蒸发掉了。这是整体规划的一部分，它创造了一个与周围环境密不可分的空间，我们相信，奉行"幼儿园是一所大房子"的初始理念，该设计可以融入周围农业社区中的那些房屋。

我们的目标是创造这样一个空间，不仅供孩子娱乐，还要通过融入周围的自然环境来创造一加一大于二的效果。

Hakusui Nursery School

This nursery school in Sakura, Chiba was planned to accommodate 60 pupils. Seiyu-Kai, a local social welfare firm specializing in elderly care facilities approached us for this project. The overarching concept for this plan started with an idea: a nursery school is a large house. Surrounded by mountains and forest, the southern area of the site rests on a gentle slope. Putting this topography to use, we designed the school room to resemble a large set of stairs.

One unique feature of this large house, for example, is that a 3-year-old child is in the same room as a 5-year-old and these children of different ages can interact in the wide, one-room space. Additionally, should a 3-year-old child be sleeping while a 5-year-old child plays nearby, these differences in rhythm rein-

项目名称：Hakusui Nursery School
地点：Sakura-shi, Chiba, Japan
建筑师：Kentaro Yamazaki
结构工程师：ASD
设施工程师：Yamada Machinery Office
平面设计：Shunpei Yokoyama Design Office
用地面积：1,046.64m²
总建筑面积：530.28m²
结构：wooden
施工时间：2014.1
竣工时间：2014.11
摄影师：©Naoomi Kurozumi (courtesy of the architect)

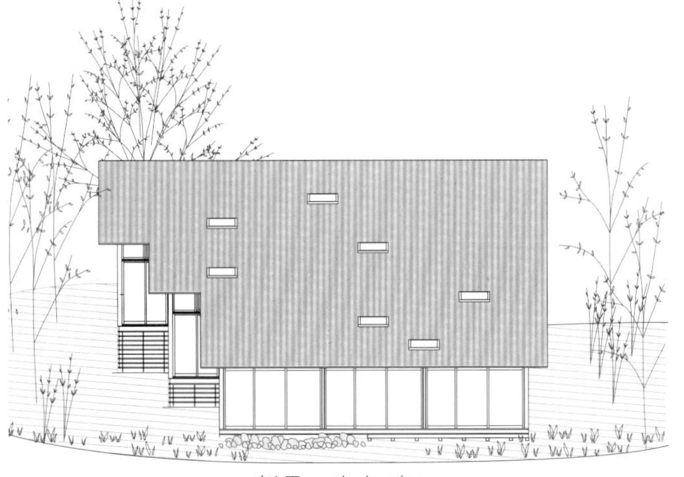

南立面 south elevation

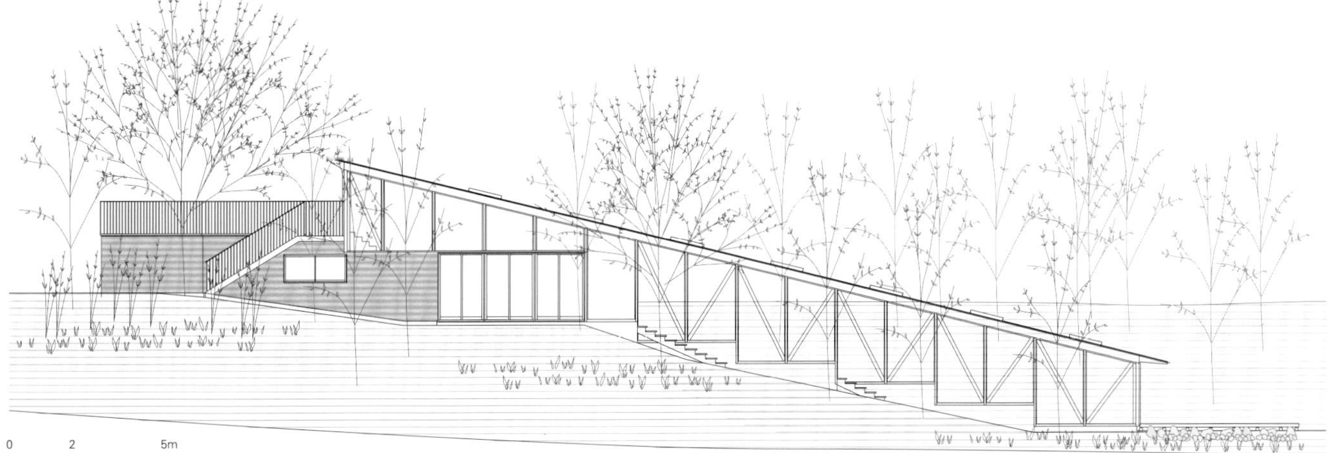

西立面 west elevation

force the "domestic" qualities of this "large house". Keeping in line with Seiyu-kai's longstanding (26-year) philosophy, we made it a priority to minimize the number of blind-spots within the room in order to keep safety measures as unintrusive as possible.
Utilizing large sliding window frames along the northern and southern faces, along with the space created by the slope, ventilation is created as a breeze draws in from the south of the structure, passes through the forest-like interior pillars making its way upwards before finally blowing out across the terrace on the northern side.
A small pond on the southern end which gathers rainwater as well as water from sprinklers installed on the southward-slanted roof also contributes to the airflow as the water is absorbed into the wind that blows into the structure. This was part of a comprehensive plan to create a space that was inseparable from the environment that surrounds it. We believe that in pursuing the original concept of a nursery school as a large house we achieved something not unlike the houses in farming communities that surround it.
Our goal was to create a space that was not only fun for the children but via blending into the nature around it, foster an experience that was greater than the sum of its elements.

Yamazaki Kentaro Design Workshop

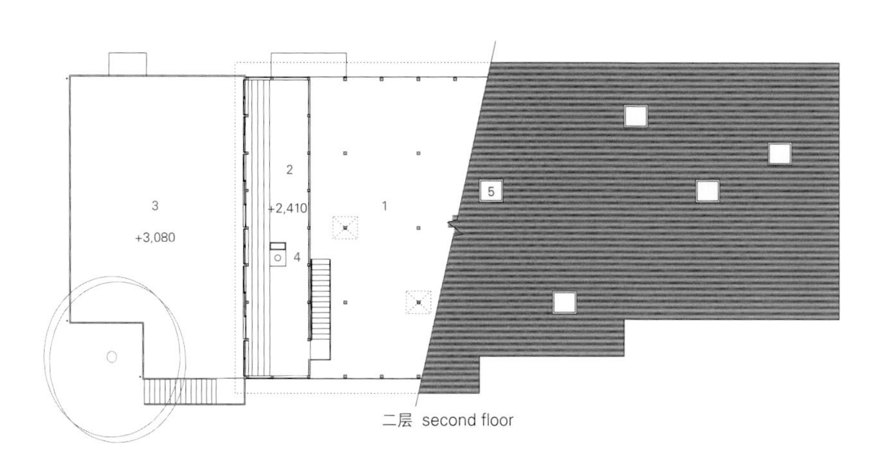

二层 second floor

1 上空空间
2 瞭望塔
3 平台
4 暖空气进气装置
5 顶部采光

1. void
2. watch tower
3. deck
4. warm air intake duct
5. top light

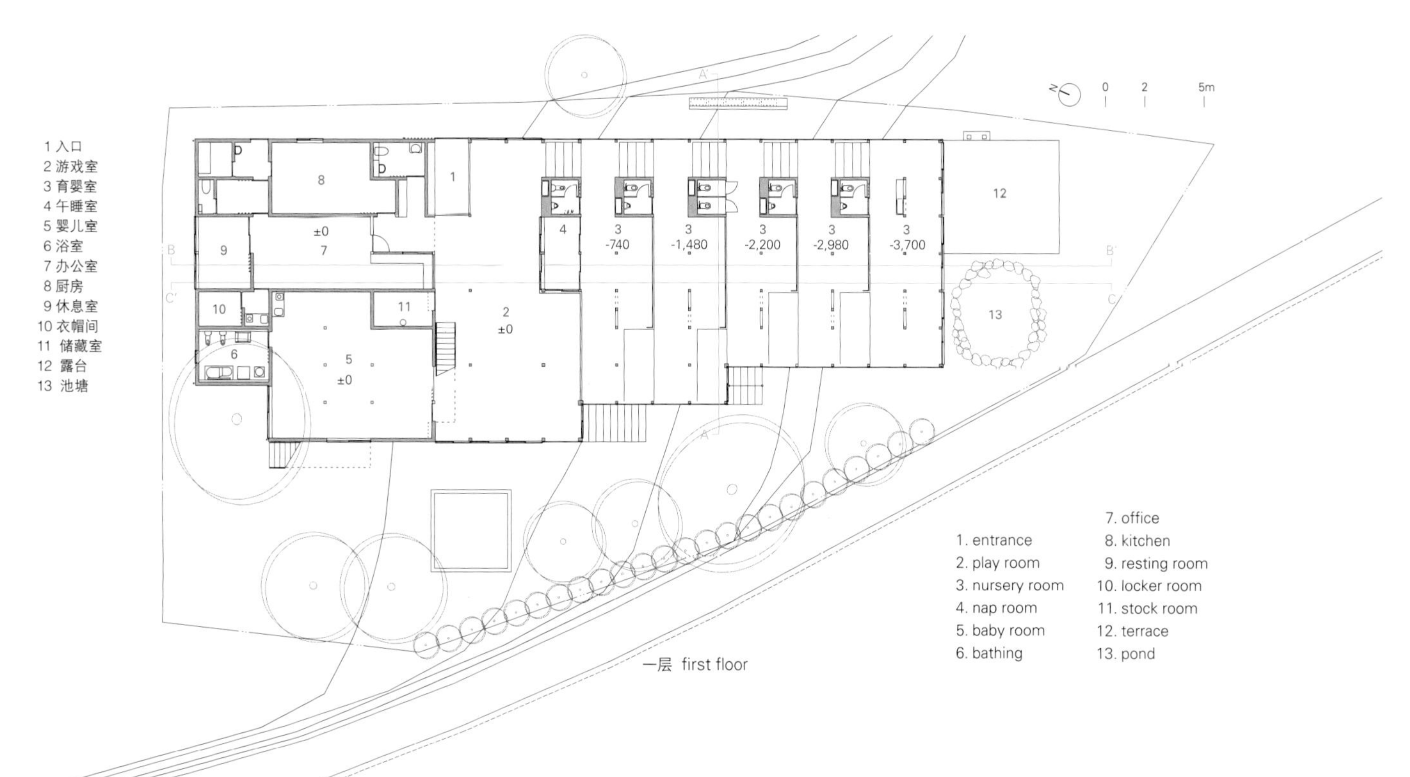

一层 first floor

1 育婴室 2 地下储藏室
1. nursery room 2. basement storage
A-A' 剖面图 section A-A'

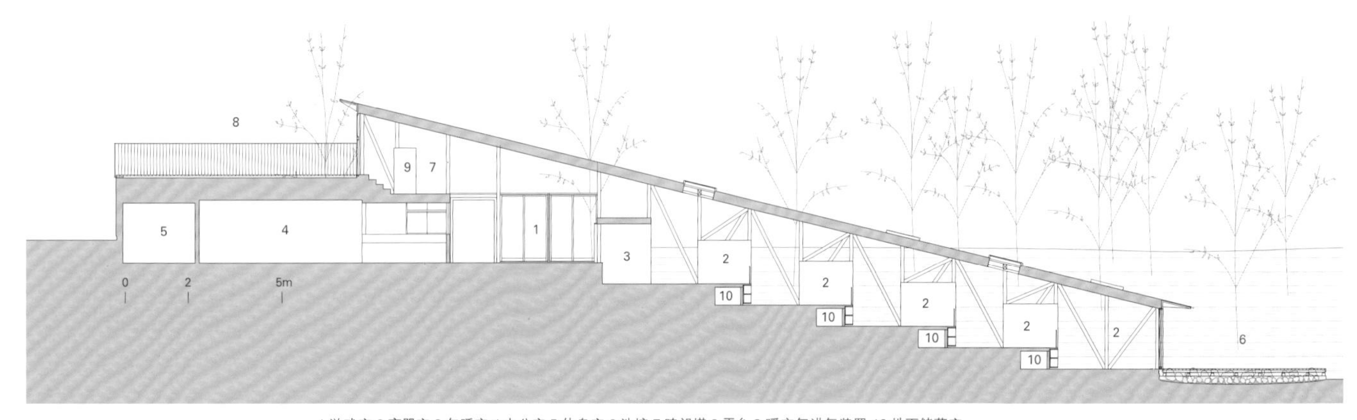

1 游戏室 2 育婴室 3 午睡室 4 办公室 5 休息室 6 池塘 7 瞭望塔 8 平台 9 暖空气进气装置 10 地下储藏室
1. play room 2. nursery room 3. nap room 4. office 5. resting room 6. pond 7. watch tower 8. deck 9. warm air intake duct 10. basement storage
B-B' 剖面图 section B-B'

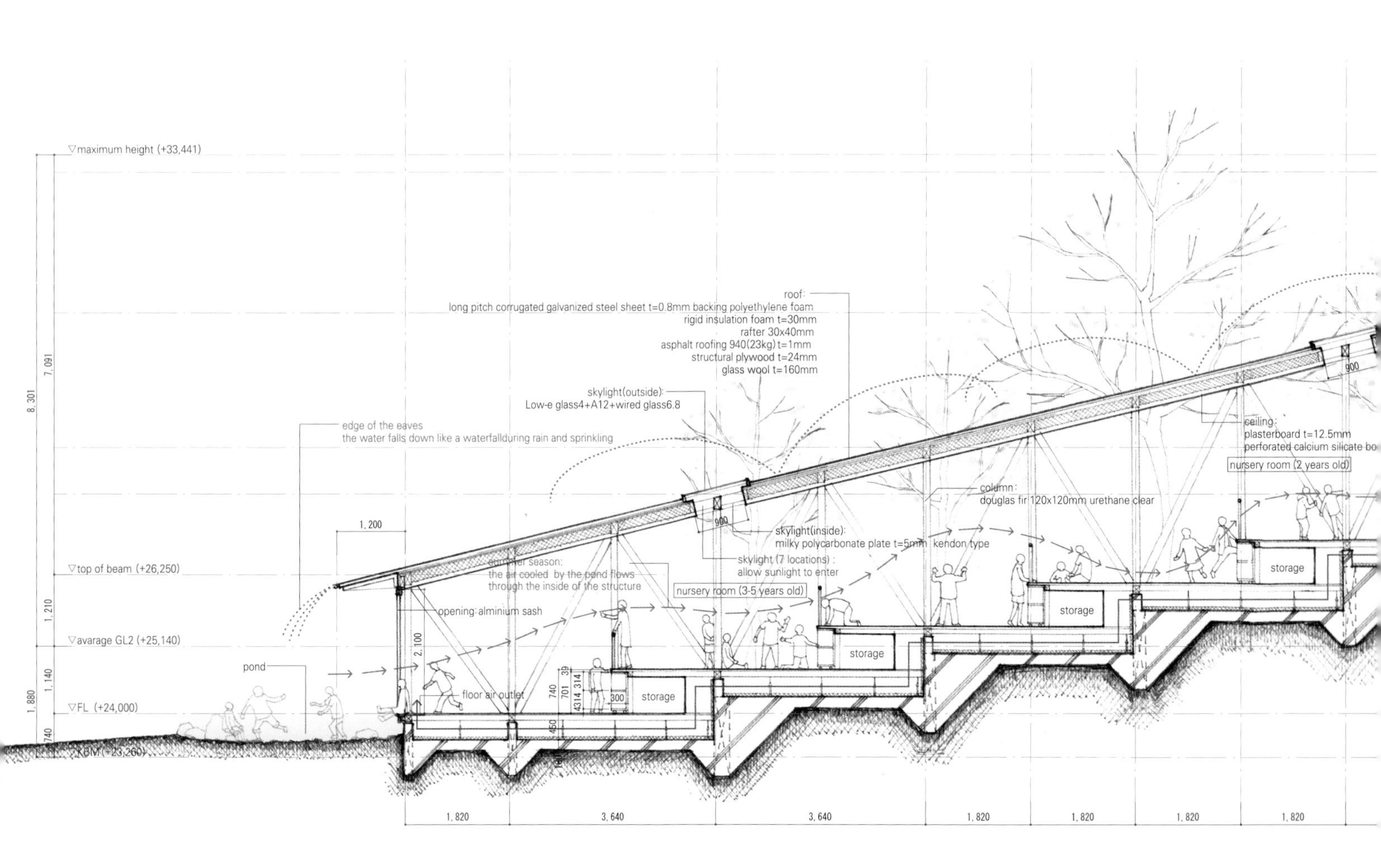

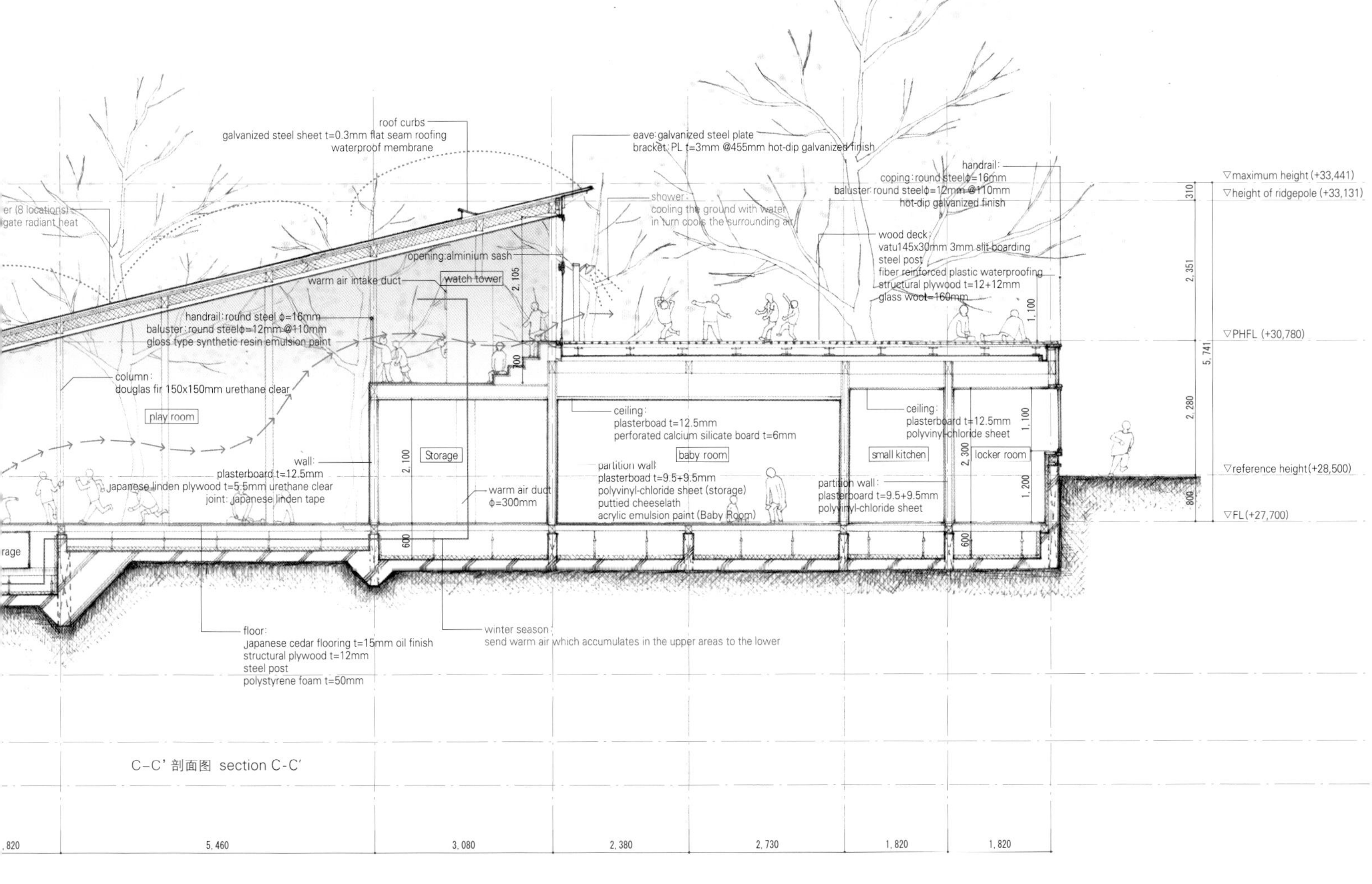

C–C’ 剖面图 section C-C’

坎普尔顿幼教中心

CO-AP

该项目敏锐地在悉尼西部一处原工业厂房场址处建造了一所可容80人的幼儿园。

根据建筑设计概要，这间可容80人的幼儿园拥有室内外的游戏空间，光线充足，通风良好，并且采用低挥发性材料。该建筑并没有采用通常幼儿园建筑的明亮的色调，而是选择了自然的颜色，材料与抛光也均属上乘，营造了一种熟悉的家的感觉。现有仓库上的一个大型切口创造了户外的娱乐空间，其余屋顶则遮蔽建筑内部为其提供荫凉。仓库的建筑元素尽可能地保留原来的状态，来揭示这个地块的发展史。

新的室内设计采用弹出式天窗可以让阳光照进游戏室。该项目仿照附近街区的城市布局，在私人领域（游戏室）的连续沿街立面设计了绿地公共空间（户外游戏室）。在设计中所采用的初级形态和柏拉图结构使孩子们拥有接受教育与开发想象力的机会。我们与景观设计师合作，整合了一系列原本就不均匀并且有点危险的隆起地面。这些空间可提供刺激、挑战平衡力和提升风险评估能力的游戏区域，是儿童早期发展的基础。

适应性再利用是这个项目可持续发展的核心。从一开始，客户就有创造一个可持续项目的想法，寻找现存工业场址是实现这一目标的第一步。我们对工业旧址适应性再利用的方法对场地干预较少，不会被资金限制，同时也是对工业遗产的尊重。

打开一部分现有屋顶可使自然光线透过并使新建筑的室内自然通风。低挥发性材料和FSC认证木材产品规格的选择增添了项目的可持续发展能力。该项目设计试图补偿能源消耗，在小一点的孩子们的空调游乐室与午睡区域采用废气净化系统。

景观绿化区和户外游戏室采用大量遮荫的树木，灌木林还有孩子们可以学习植物知识与健康饮食的厨房。模块化系统建造出了功能性好而又经济实用的建筑，使施工和装修可在12周内完成。

我们的第一家幼教中心建造项目抓住机会，对当前商业模型的风险进行了评估，并阐述了该模式替代早期模式的理由。如今，人们对于幼儿园建筑的要求越来越高，坎普尔顿幼教中心希望该模式可以振兴这一建筑类型。

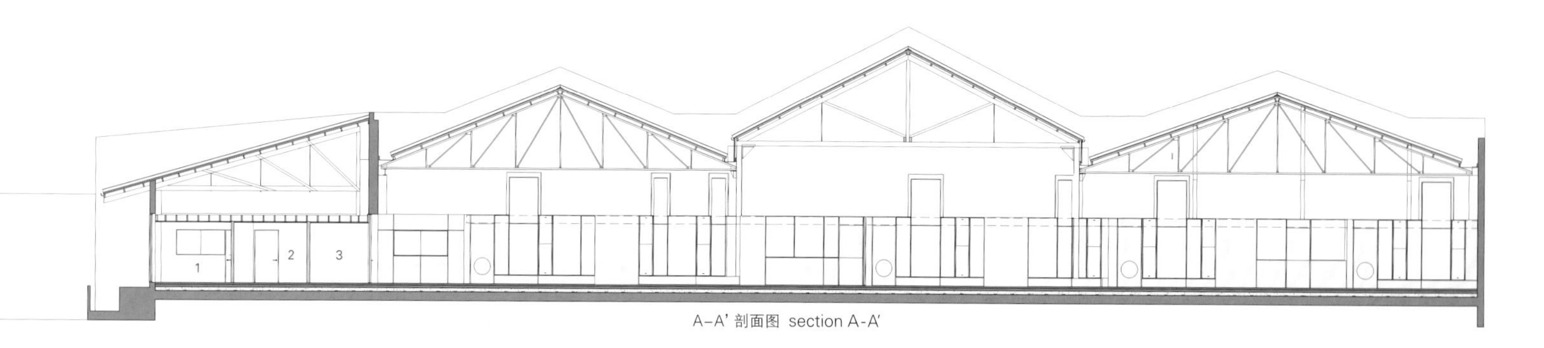

A-A' 剖面图 section A-A'

1 办公室
2 储藏室
3 卫生间
4 花园棚屋
5 厨房
6 两到五岁儿童户外教室
7 两到三岁儿童室内教室

1. office
2. store
3. WC
4. garden shed
5. kitchen
6. 2~5-year-old outdoor classroom
7. 2~3-year-old indoor classroom

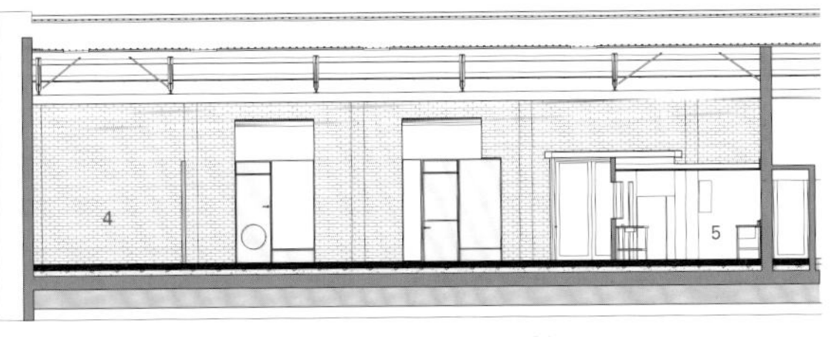

B-B' 剖面图 section B-B'

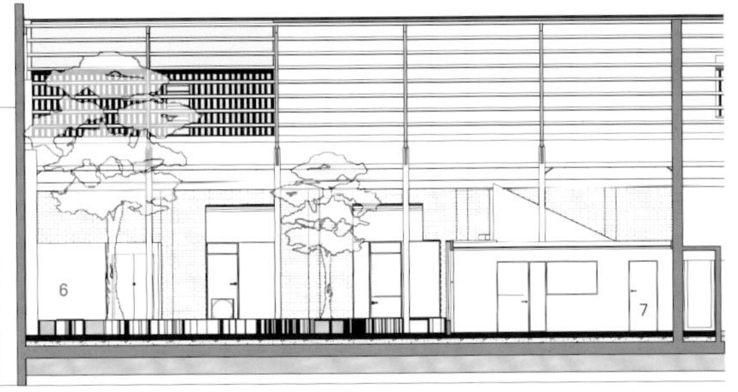

C-C' 剖面图 section C-C'

Camperdown Childcare Center

This project sensitively inserts new architecture into a former industrial warehouse building, located in the inner west of Sydney, for a new 80-place childcare center.

The brief called for a new 80-place childcare center that would accommodate indoor and outdoor playspaces, an abundance of daylight and natural ventilation and a low VOC materials specification. Dominant bright colors, usually associated with institutional childcare accommodation, were avoided in preference to a palette of natural, honest materials and finishes to create a familiar residential feel. A large cut-out to the existing warehouse roof created outdoor playspaces open to the elements while the remaining roof provides shade and shelter over the new internal building forms. Where possible, the warehouse building elements retained were left in original condition to reveal layers of site history.

The new interior architecture incorporates pop-up clerestory windows allowing sunshine to penetrate deep into the playrooms. The plan is arranged much like the urban fabric of the neighborhood, where a continuous street facade of the private realm (playrooms) addresses the communal space of the village green (outdoor playspaces). The use of primary shapes and platonic forms in the building elements give opportunities for imagination and education. In collaboration with the landscape architect we integrated a series of naturally non-uniform and adventurous elevated spaces amongst prerequisite soft-fall surrounds. These provide stimulating play areas

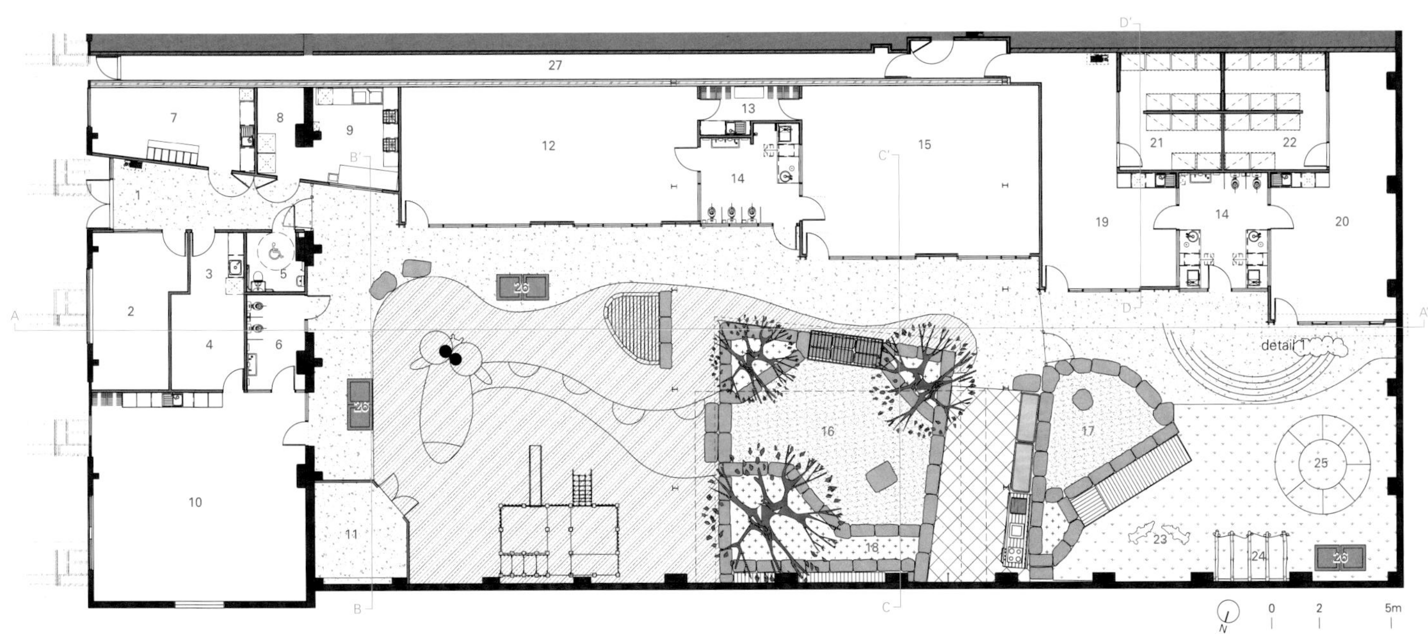

1 门厅 2 办公室 3 洗衣房 4 储藏室 5 残障人士卫生间 6 卫生间 7 教工房间 8 餐具室 9 厨房 10 四到五岁儿童教室（20名） 11 花园棚屋 12 三到四岁儿童教室（20名） 13 储藏室及工艺室 14 卫生间及换尿布室 15 两到三岁儿童教室（16名） 16 沙坑（两到五岁儿童） 17 沙坑（两岁以下婴幼儿） 18 花床 19 一岁以内婴儿教室（12名） 20 一到两岁婴幼儿教室（12名） 21 一岁以内婴儿卧室 22 一到两岁婴幼儿卧室 23 木头动物玩具 24 木质杂物间 25 球坑 26 花槽 27 公共区域/防火走廊

1. lobby 2. office 3. laundry 4. store 5. accessible WC 6. WC 7. staff room 8. pantry 9. kitchen 10. 4-5-year-old classroom (20 children) 11. garden shed 12. 3-4-year-old classroom (20 children) 13. store & craft 14. WC & nappy change 15. 2-3-year-old classroom (16 children) 16. sand pit (2-5 year old) 17. sand pit (0-2 year old) 18. garden bed 19. 0-1-year-old classroom (12 children) 20. 1-2-year-old classroom (12 children) 21. 0-1-year-old cot room 22. 1-2-year-old cot room 23. timber animal 24. timber cubby 25. ball crater 26. planter 27. common area / fire corridor

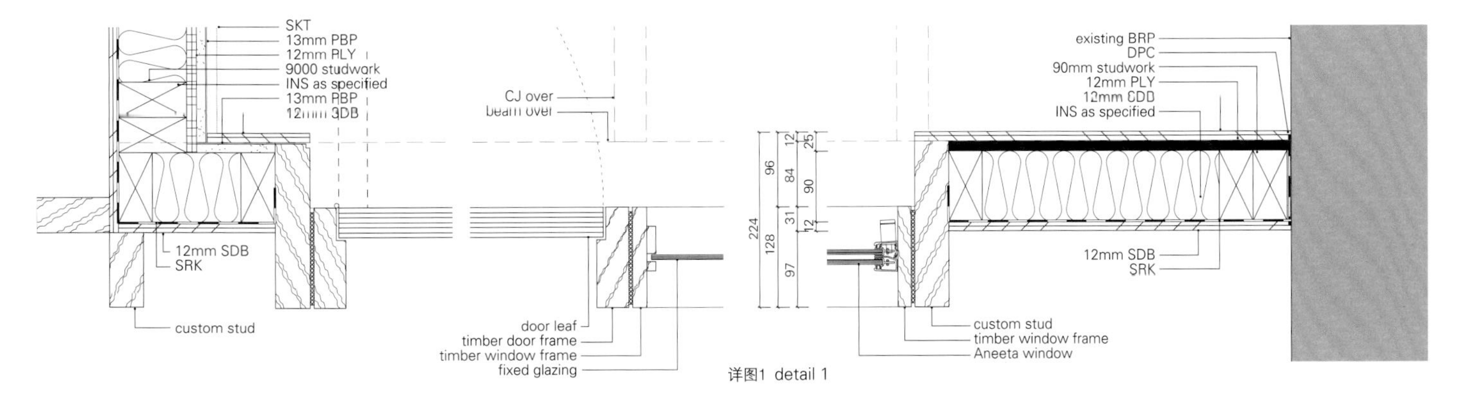

详图1 detail 1

to challenge the balance and promote the risk evaluation that are fundamental for early childhood development.

Adaptive re-use is at the core of the approach to sustainability for this project. From the outset, the client had a vision for creating a sustainable project and finding the existing industrial site was the first step of achieving the goal. Our approach to adaptive re-use of the industrial building was for light intervention which was a result of both budget constraints and an appreciation of industrial heritage.

Opening up a section of existing roof provided opportunities for natural light and natural ventilation to permeate the new interior architecture. The selection of low VOC materials and finishes and specification of FSC certified timber products add to the sustainability credentials of the project. This strategy attempts to offset the energy consumed as a consequence of DECS requirements for conditioned playrooms and sleep areas for the younger children.

The approach to landscaping and outdoor playspace is for abundant planting of shade trees, bushes, and a kitchen garden for the children to learn about plants and healthy eating.

A modularized system of fabrication yielded an economical and efficient building, allowing for on-site construction and fit-out to be completed in 12 weeks.

Being our first childcare center project, we have used the opportunity to evaluate current risk-adverse commercial models and reassess superseded early childhood rationales. Camperdown Childcare Center hopefully contributes to invigorating this increasingly demanding building typology. CO-AP

项目名称：Camperdown Childcare Center
地点：58 Denison Street, Camperdown NSW 2050, Australia
建筑师：CO-AP / 项目团队：Will Fung, Patrik Braun, Tina Engelen
市政府：Marrickville Council
景观设计师/游戏区顾问：Fiona Robbé Landscape Architecture
结构设计：Partridge Structural / 水力学：Glenn Haig & Partners
暴雨设计：Ecological Design / 声效：Acoustic Logic / BCA：BCA Logic
机械：Northrop Consulting Engineers
承包商：builder_Blitz Group Pty Ltd / landscape_Jamie Miller Landscapes / mechanical_R & J Airconditioning Services
客户：Explore & Develop Camperdown
总建筑面积：1,113m² (outdoor play space_588m²)
造价：USD 850,000 / 项目持续时间：15 months (3 month build)
竣工时间：2014.4
摄影师：©Ross Honeysett (courtesy of the architect)

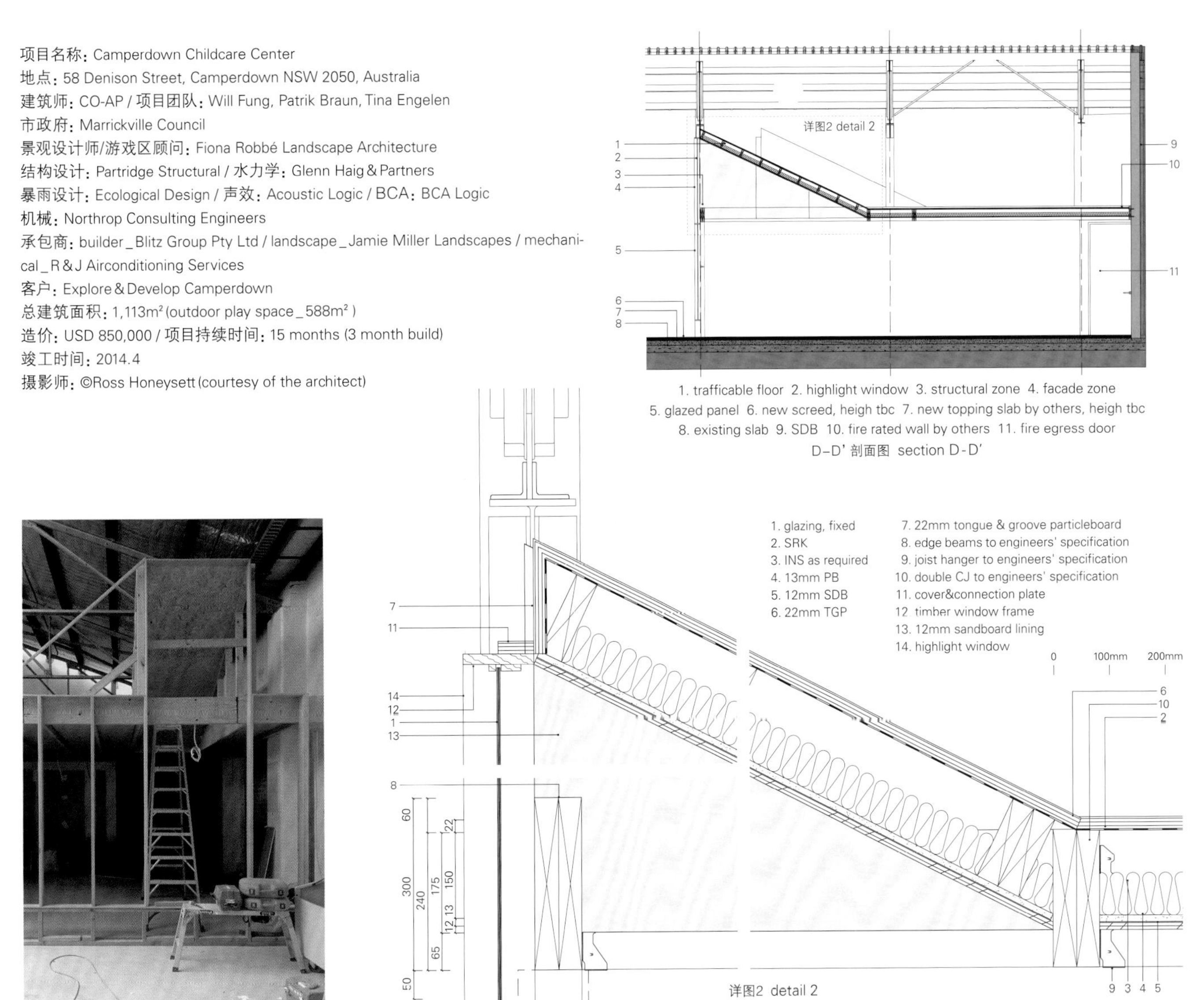

1. trafficable floor 2. highlight window 3. structural zone 4. facade zone 5. glazed panel 6. new screed, heigh tbc 7. new topping slab by others, heigh tbc 8. existing slab 9. SDB 10. fire rated wall by others 11. fire egress door

D-D' 剖面图 section D-D'

1. glazing, fixed
2. SRK
3. INS as required
4. 13mm PB
5. 12mm SDB
6. 22mm TGP
7. 22mm tongue & groove particleboard
8. edge beams to engineers' specification
9. joist hanger to engineers' specification
10. double CJ to engineers' specification
11. cover&connection plate
12. timber window frame
13. 12mm sandboard lining
14. highlight window

详图2 detail 2

“花生”幼儿园

UID Architects

建筑是遮风挡雨的容身之所，人们在陆地上筑屋建瓴，并使之有别于自然环境。然而，在我看来，最原始的环境可以影响人类与生俱来的动物本能，也就是说，人们在自然环境中发现一个栖身之所，不论环境好坏与否都能随遇而安，这种人与自然密不可分的关系是一种自然的状态。

然而，大环境每时每刻都在改变。天地万物不断变化，彼此联系，人类也逐步发展，五种感官变得敏锐，心智得到开拓。在这种情况下，面向新生儿的“花生”幼儿园可以说是首次尝试，这家幼儿园提供了不同的自然环境，让建筑促进婴儿感官的发展。

室外部分被称作探索森林，紧挨着护理中心，只要时间和天气允许，孩子们就能到室外活动。有时候自然环境稳定舒适，有时候比较恶劣。我希望在这里孩子们能通过与自然的互动适应各种不同的环境。圆形有很强的向心力，是一个完结的图形。但是两个圆形重叠起来，形式就发生了变化，具有多个矢量，失去向心力。花生幼儿园不是被划分成不同区域的空间，而是与环境的自然联系。我希望孩子们能在这里发展各种感官，因为“花生”不是一个完结的图形。

幼儿园天花板最低高度是2.1m，因此从功能角度讲，护理人员不会感到不便。我还考虑过空间大小，例如，宽阔的水平空间顺着坡形地板延伸至平滑的天花板。我认为养育孩子的地方要接近自然，所以人与自然的亲密接触就是一种乐趣。

孩子的空间被特别地设计成平缓的斜坡，在那里，孩子们可以感受到从叶子的缝隙间透过的阳光、自然中的生灵、四季的花香、淅沥的雨声。育儿室外面环绕着的半封闭长廊，有一些缓坡，可以让人漫步其中，这是孩子们探索的乐园，是森林画廊，准妈妈和妈妈们也可以在这里休闲放松。我比较注重目光交流中体现的信任感。我采用了“一屋

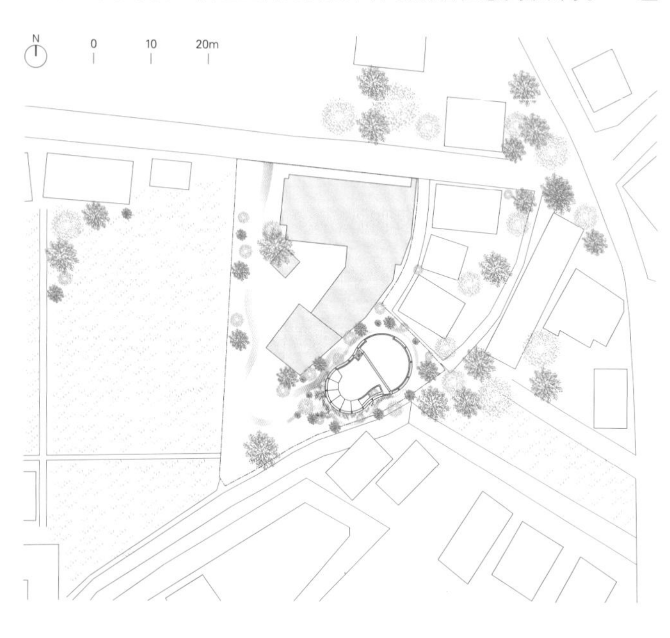

1 入口 2 育婴室 3 更衣室 4 婴儿食品室 5 办公室 6 储藏室 7 浴室 8 探索森林 9 走廊
1. entrance 2. baby room 3. locker 4. formula room 5. office 6. storage 7. bathing room 8. forest of the exploration 9. gallery

式"的格局，但是明确划分了管理区、儿童房和画廊。护理人员很容易看到孩子，管理区桌子的高度和儿童房的地板高度一致，以确保安全。另外，考虑到婴儿是通过观察学习的，所以这种目光交流变得很自然。该设计让老师在管理区可以环视整个房间，这样就能提高空间的功能性。通过建立孩子与护理人员、建筑与自然、人与自然环境之间的平等关系，可以创造丰富的环境。

Peanuts

Architecture is a shelter to protect the body from nature Therefore, it takes root in the land but tries to be distinct from the natural environment. However, I think that the environment working on humans' inherent animal instincts is the primary state. In other words, the relationship that exists in the environment without being divided each other while one discovers a place to stay in the natural environment and accept good and strict environmental points is a natural state. Yet the entire environment continuously changes every moment. All of creation, feeling that an immutable thing does not exist in the world, connects the living beings and contrib-

utes to the formation of humans who can sharpen the five senses and open their minds. In this context, it can be said that this "Peanut" is a first trial as a preschool functioning for newborns. The space produces a variety of natural environments, engaging the architecture to enhance the sensitivity of impressionable babies.

The outside space, called the forest of exploration, is connected to the childcare space, and babies can explore the outside grounds depending on the time and weather. Sometimes the natural environment provides stable and comfortable conditions; the other times, it provides severe conditions. I hope that it could become a space that could take in living beings in a variety of environments by interacting with nature. One circle is the completed form because it has strong centripetal characteristics. However, two overlapping circles change the form into that is not centripetal with various vectors. It becomes the space that does not prescribe the domain as various places to stay, and it is connected gently to the envi-

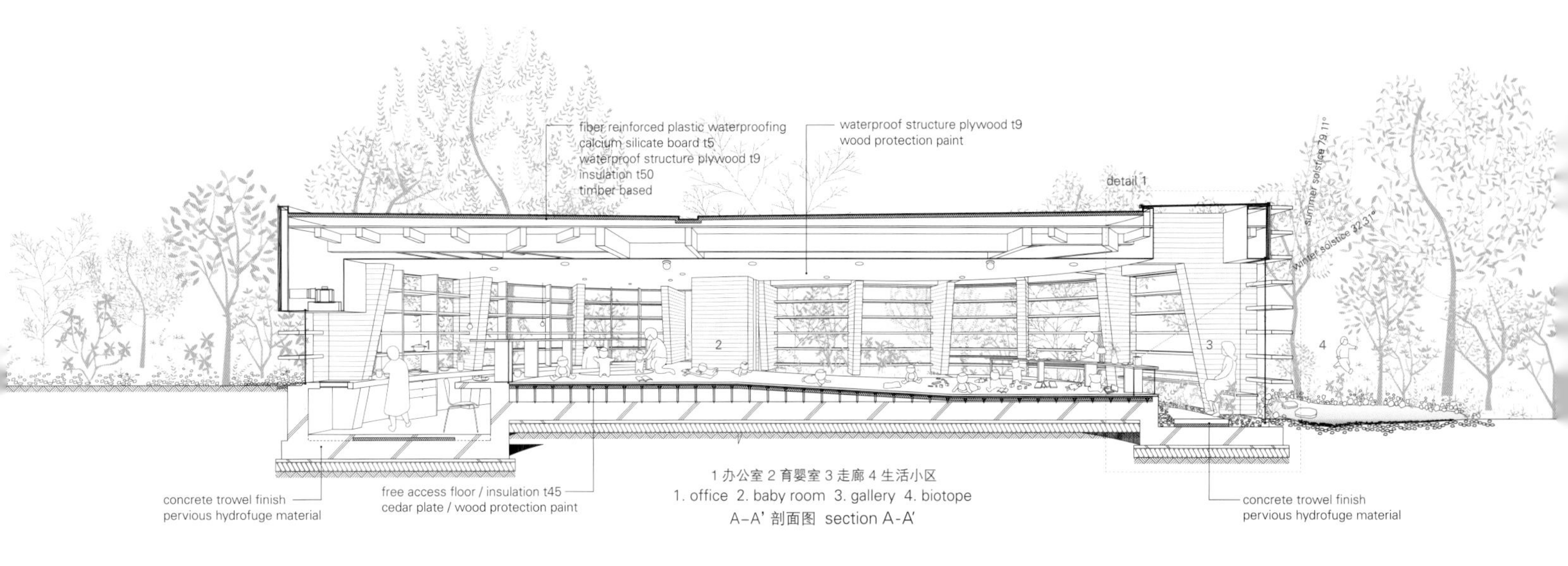

1 办公室 2 育婴室 3 走廊 4 生活小区
1. office 2. baby room 3. gallery 4. biotope
A–A' 剖面图 section A-A'

项目名称：Peanuts
地点：Kinosho-cho, Fukuyama, Japan
建筑师：UID Architects
顾问：Konishi Structural Engineers, Toshiya Ogino Environment Design Office
总承包商：Home Co. Ltd
结构系统：timber structure
用地面积：1,501.81m² / 建筑面积：118.74m² / 总楼面面积：118.74m²
材料：
exterior _ cider plate, wood protection paint,
interior _ structural plywood, exposed concrete, wood protection paint, cherry flooring
竣工时间：2012.3
摄影师：©Hiroshi Ueda (courtesy of the architect)

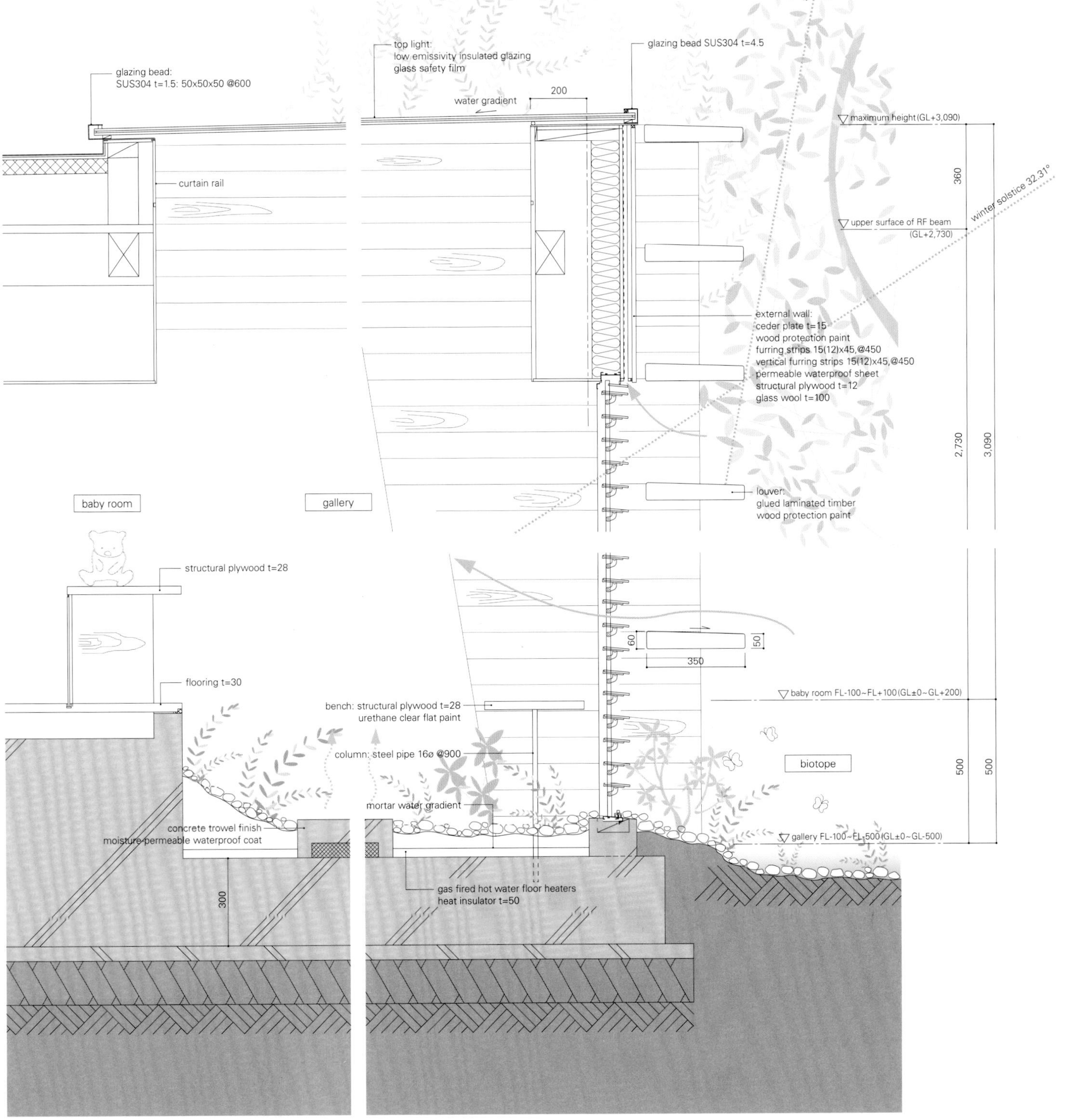

summer solstice 79.11°
winter solstice 32.31°
glazing bead: SUS304 t=1.5: 50x50x50 @600
top light: low emissivity insulated glazing glass safety film
glazing bead SUS304 t=4.5
water gradient
200
maximum height (GL+3,090)
upper surface of RF beam (GL+2,730)
curtain rail
360
external wall:
ceder plate t=15
wood protection paint
furring strips 15(12)x45,@450
vertical furring strips 15(12)x45,@450
permeable waterproof sheet
structural plywood t=12
glass wool t=100
2,730
3,090
baby room
gallery
louver:
glued laminated timber
wood protection paint
structural plywood t=28
60
50
350
flooring t=30
baby room FL-100~FL+100 (GL±0~GL+200)
bench: structural plywood t=28
urethane clear flat paint
column: steel pipe 16ø @900
biotope
mortar water gradient
500
500
concrete trowel finish
moisture-permeable waterproof coat
gallery FL-100~FL-500 (GL±0~GL-500)
300
gas fired hot water floor heaters
heat insulator t=50

详图1 detail 1

ronment. I expect that babies can develop sensitivity in this architecture because "Peanut" is not a completed form. The lowest height of the ceiling in this building is 2,100mm, so it is not inconvenient for childcare providers in terms of the functional aspects. I thought about the scales as well. For example, the expanses of the level spaces change based on the slope of the floor to a flat ceiling overall. Because I think that contact with nature is a place for raising humans, contact with nature becomes a pleasure.

Specifically, the babies' space has the incline such as the gentle topography, they enjoy feeling sunshine filtering through foliage, side fluctuation, creature in nature, fragrant in season, and the rainy sound etc. And the semi-outdoor space having some slants and excursion characteristics surrounds a baby room with variety: a place of the exploration for children and "a gallery of the forest" can relax for pregnant women

and mother. I value confidence to join each other together from the relations of the eye level. I adopted one-room style while I zoned administrative parts and babies' room and gallery definitely. The glance of a childcare person and the baby becomes near, and security occurs by making desk height in the administration the same as the height of the floor of the babies' room. In addition, judging from the glance of "learning from seeing" between babies, it becomes the natural relations. I intend to raise functionality by being able to look around the whole room on the administrative side. I create rich environment by joining together for equivalent relations with babies and child minders, building and nature, person and natural environment. UID Architects

巴黎日托中心

Rh+ Architecture

在巴黎南部，RH+建筑师事务所建造了一座新的日托中心。该项目位于一条弯曲的小路上，周围都是工业时期遗留下来的典型的法式历史建筑。附近稀稀落落的屋舍、蜿蜒的路街、2.5m的高差，使人感觉置身于小城镇。事实上，日托中心的位置受到了古斯塔夫·杰弗里大街弯道的限制，要想与城市格局和谐相融，就必须利用这些限制。

该项目的主要理念是设计一栋既适合城市环境又别具一格的现代简约建筑。同时，还能凸显周围建筑的历史感，创造一种全新的都市风格。因此，为了使建筑外观适应道路的高度差，减少坡度影响，产生整体统一的效果，该建筑被放置在一个基座上。一系列空间错落有致地排列在基座上面，使人对建筑产生不同的感觉和认识，并在视觉上与街道保持联系。

该项目具有家庭规模，其上方连续的体量与其他建筑的规模相适应。北边的体量最高，紧贴着一座五层楼，这栋楼也是街道中最重要的建筑。中间的部分是日托中心的入口，也是最长的。三座体量都使用白色混凝土、斜屋顶，与周边传统建筑环境相呼应。窗口安装了中央内置式气窗，装有儿童固定窗户（最高的部分为1.3m）。

立面材料采用了混凝土立体造型墙砖，墙砖的花纹经过多次等比例试验制成，这种立面同上方房间光滑的表面形成了强烈的对比，并在阳光的照射下产生了无穷的变化。立面部分与一层相结合，又突出了基座的概念。

托儿所应该是孩子们认识世界的地方，在互相尊重、安全、舒适的环境中，孩子们在互动中学习。因此，这家日托中心采用了大量的玻璃隔断划分不同的区域，以获得最佳的自然采光。每个区域都使用不同的颜色，让空间变得有趣而生动。游戏室面向花园，外部安装了百叶窗避免太阳直射。

对孩子智力的刺激作用并不是以线性的方式发生的，教育环境越丰富，对孩子们来说就越有利。本项目室内空间的设计理念正是来源于这种想法，设计师认为建筑应该有助于激发孩子们的感官。另外，窗

户的接缝和窗框使用铝和木材，并且涂有颜色。另外，庭院和阳台上都铺着印有彩色圆圈图案的橡胶软垫。

Day Care Center in Paris

The project is located in the south of Paris, in a small curved street, surrounded by historical and typical French buildings of the industrial period. The low density of the neighborhood, the street shape in addition to a 2.5m difference in height give the impression of a small provincial town. In fact, the curve of the Gustave Geoffroy street gives this exceptional site several constraints that have to be used to generate a well-integrated project to its urban fabric.

The main ideas of this project are to design a building that perfectly fits in its heterogeneous urban surroundings, and also to provide a contemporary and simple architecture, yet discrete and remarkable. This will, in the same time, highlight the history of the buildings nearby, and create a new urbanity. In this way, the building has been designed on a base, to mark an urban facade that accompany the difference in height of the street, reduce the slope effect and give a certain unity to the whole. On this base, a set of volumes have been placed in such a way that it allows different perceptions and perspectives and also preserves a visual connection to the street.

This project has a domestic scale, the succession of the volumes above it, making it correspond with the scale of other buildings. The northern volume is the highest, and it leans on

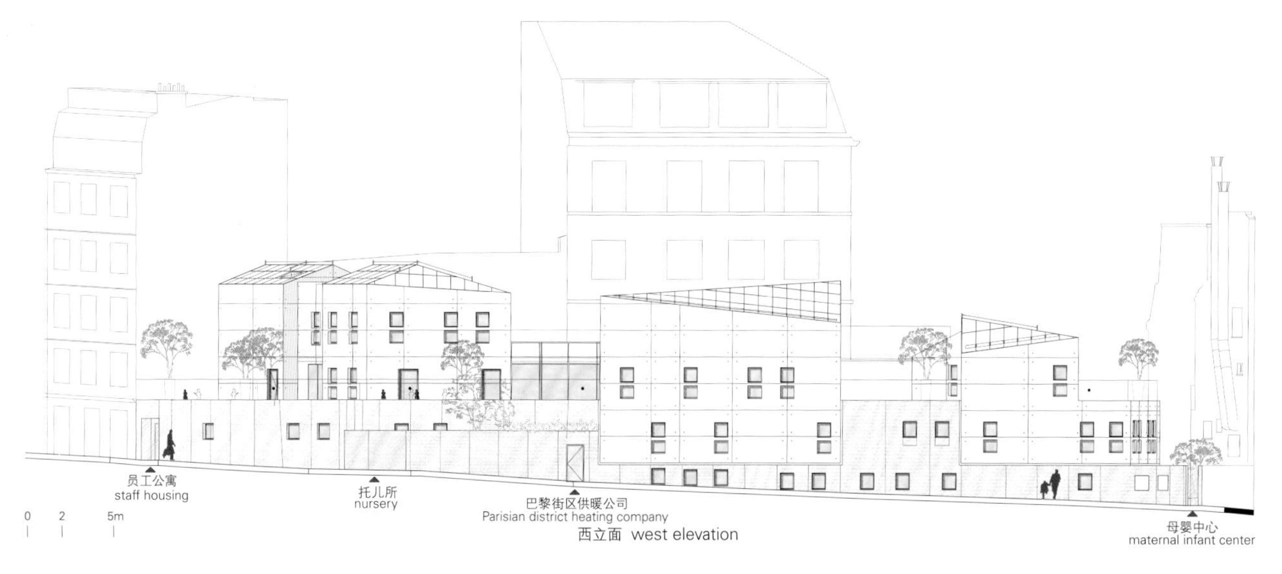

西立面 west elevation

东立面 east elevation

11
13

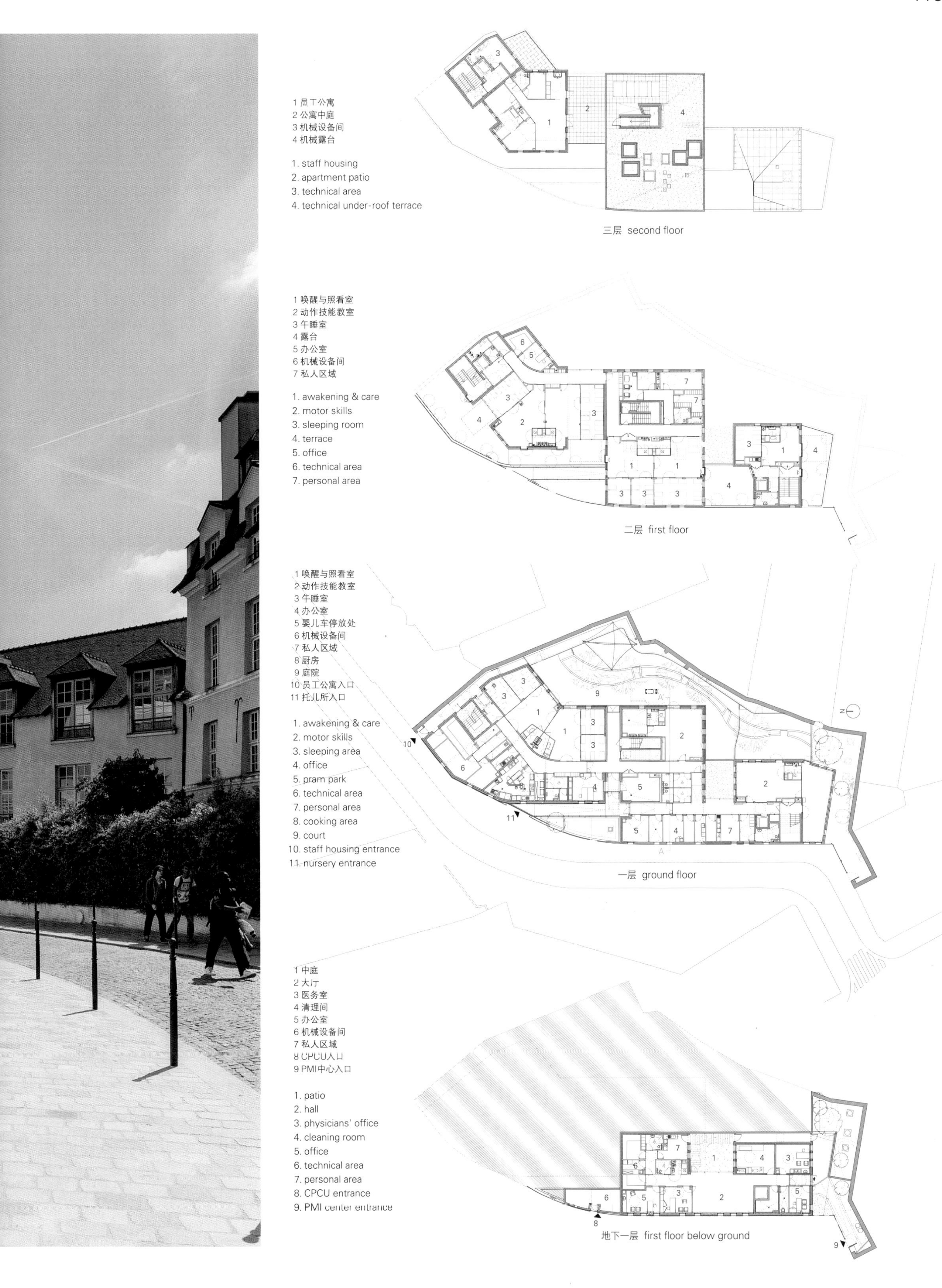

1 员工公寓
2 公寓中庭
3 机械设备间
4 机械露台

1. staff housing
2. apartment patio
3. technical area
4. technical under-roof terrace

三层 second floor

1 唤醒与照看室
2 动作技能教室
3 午睡室
4 露台
5 办公室
6 机械设备间
7 私人区域

1. awakening & care
2. motor skills
3. sleeping room
4. terrace
5. office
6. technical area
7. personal area

二层 first floor

1 唤醒与照看室
2 动作技能教室
3 午睡室
4 办公室
5 婴儿车停放处
6 机械设备间
7 私人区域
8 厨房
9 庭院
10 员工公寓入口
11 托儿所入口

1. awakening & care
2. motor skills
3. sleeping area
4. office
5. pram park
6. technical area
7. personal area
8. cooking area
9. court
10. staff housing entrance
11. nursery entrance

一层 ground floor

1 中庭
2 大厅
3 医务室
4 清理间
5 办公室
6 机械设备间
7 私人区域
8 CPCU入口
9 PMI中心入口

1. patio
2. hall
3. physicians' office
4. cleaning room
5. office
6. technical area
7. personal area
8. CPCU entrance
9. PMI center entrance

地下一层 first floor below ground

the party wall of a five-story building which is also the most important in the street. The central volume is the longest, it is also the building entrance to the childcare.

The three volumes are made of white concrete and have angled roofs that remind the traditional architectural context nearby. Openings are treated with a central built-in transom, which offers fixed windows for children (the highest part are at 1.30m).

The material facade used is the Ductal concrete. The pattern has been designed in the office after many tries in real scale, to create a material that interacts with the sun and evaluated with the changing light, but also contrasts with the smoothness of the material facade of the volumes above. It unites the ground floor and reinforce the idea of the base.

A childcare should be a place where children can develop their perception of life, and learning by stimulation. By giving a respectful, safe and comfortable environment. Thus, this project has optimal natural lighting, with a maximum of glazed modules between the different areas. Each of these areas has been treated with a different color, to make the spaces fun and animated. The playroom are widely opened to the garden, and outer blinds protect them from the sun.

The stimulation of children is not only developed in a linear way, that is, the richer the educational environment is, the better it is for the children. The project idea to design indoor spaces, comes from the thought that architecture should help stimulate the youngest senses. Besides, all the windows joints and frames are made in aluminum/wooden, lacquered in colors. The courtyard, and the children terrace are covered in a cushioned rubber floor with colored circles.

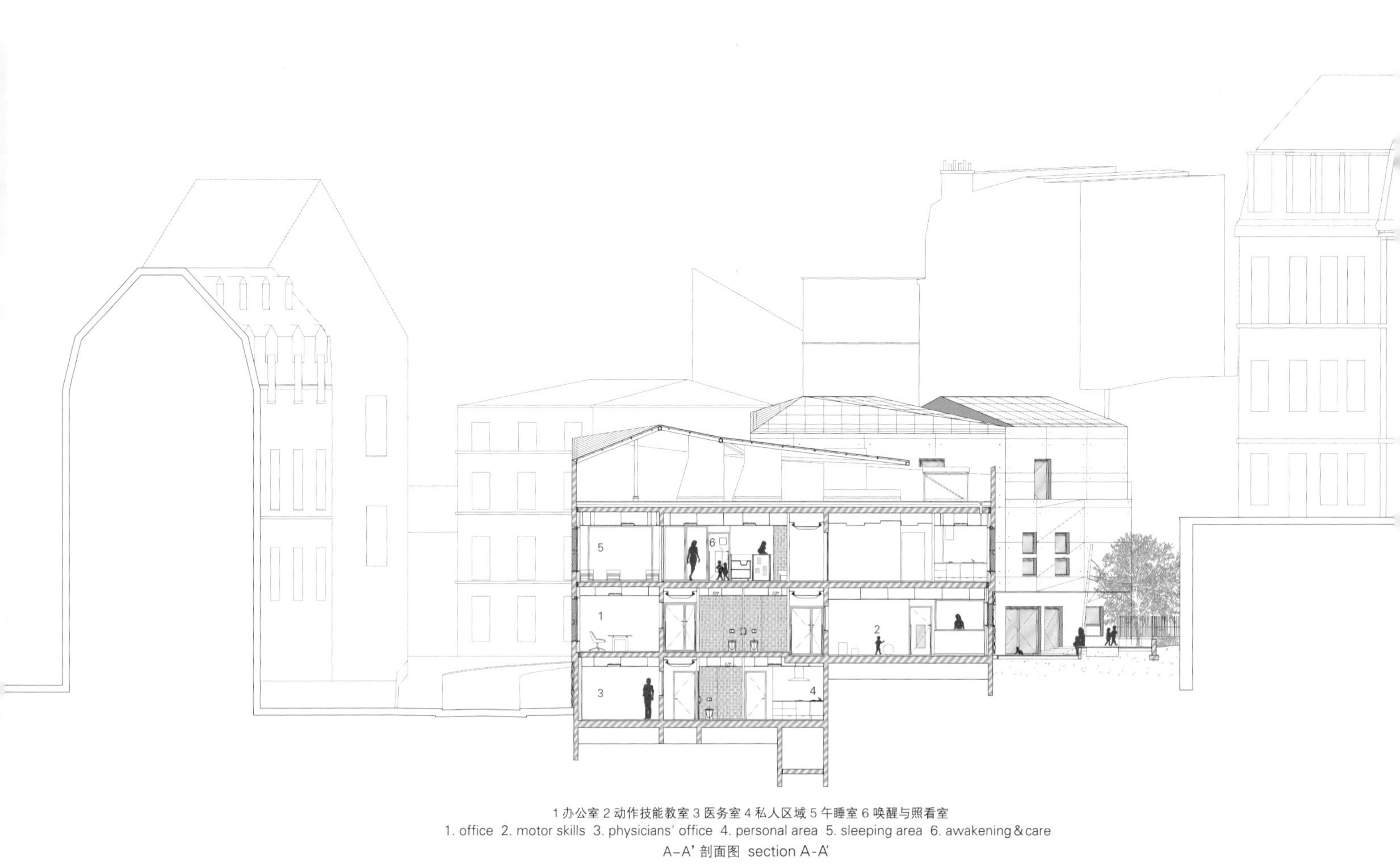

1 办公室 2 动作技能教室 3 医务室 4 私人区域 5 午睡室 6 唤醒与照看室
1. office 2. motor skills 3. physicians' office 4. personal area 5. sleeping area 6. awakening & care
A–A' 剖面图 section A-A'

项目名称：Day Care Center in Paris
地点：17 Gustave Geffroy street, Paris 13th arrondissement, France
建筑师：Rh+ Architecture
承包商：Capaldi Construction
造价师：Betom
HQE工程师：Cap Terre
客户：Ville de Paris
功能：day nursery of 66 cribs, mother and child welfare center, company accommodation
用地面积：1,300m²
建筑面积：1,800m²
造价：EUR 5,400,000
设计时间：2012
施工时间：2014
竣工时间：2014.3
摄影师：©Luc Boegly (courtesy of the architect)

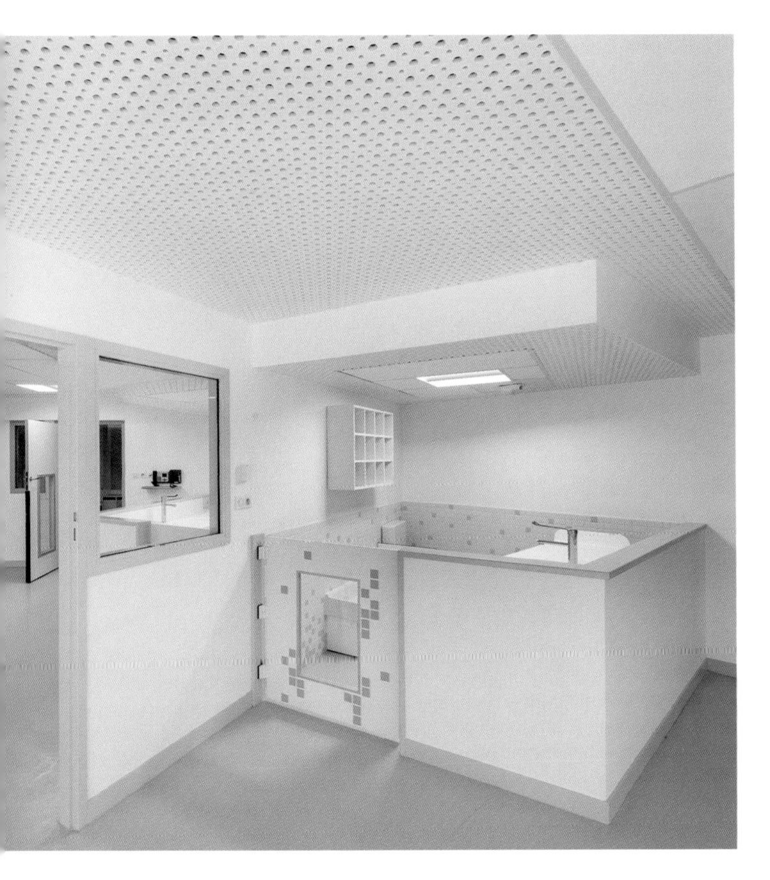

圣吉尔小学

Lens° Ass Architecten

该街景显示校舍是由两栋建筑构成的。这一点显示在立面的双重设计上。细长的窗户突出了街道的垂直节奏。该建筑物很自然地融入了周围略显正式的建筑风格。

该建筑不缺乏象征含义。正前方的树叶图案和绿色的设备罩暗示生长和发展。教室是根据年龄安排的，每层三个年龄段，教导处和校长办公室在顶楼。为了最大限度地创造开放空间，教学楼被设计得精巧紧凑，尽可能节省空间。操场适合嬉戏活动。后院的一所旧屋也能进行户外活动。"露天剧场"适合于各种学校活动。

尽管使用了混凝土、混凝土块、钢铁和玻璃等经久耐用的牢固材料，圣吉尔小学仍然营造了轻松愉快的学习环境。

Sint-Gillis Primary School

The streetscape shows a school building that takes up two building lots. This is reflected in the dual building design of the facade. The elongated windows accentuate the vertical rhythm of the street. The building integrates easily into the formal vocabulary of its surroundings.

东立面 east elevation

西立面 west elevation

The building is not devoid of symbolism. Leaf motifs on its front side and green fixture coverings allude to growth and development. The classrooms are organized according to age: three per floor. The administration and director's office are situated on the top floor. The architecture is meant to be compact and space-saving in order to create as much open space as possible. The playground invites active playing. A dilapidated building situated in the back functions as a covered outside space. The outhouse in this building volume reflects the profile of its covering on a smaller scale. The "open air theater" is used for a variety of school activities.

Despite the presence of unmovable and durable materials such as concrete blocks, steel and glass, the school presents itself as a particularly familiar and child-friendly learning environment.

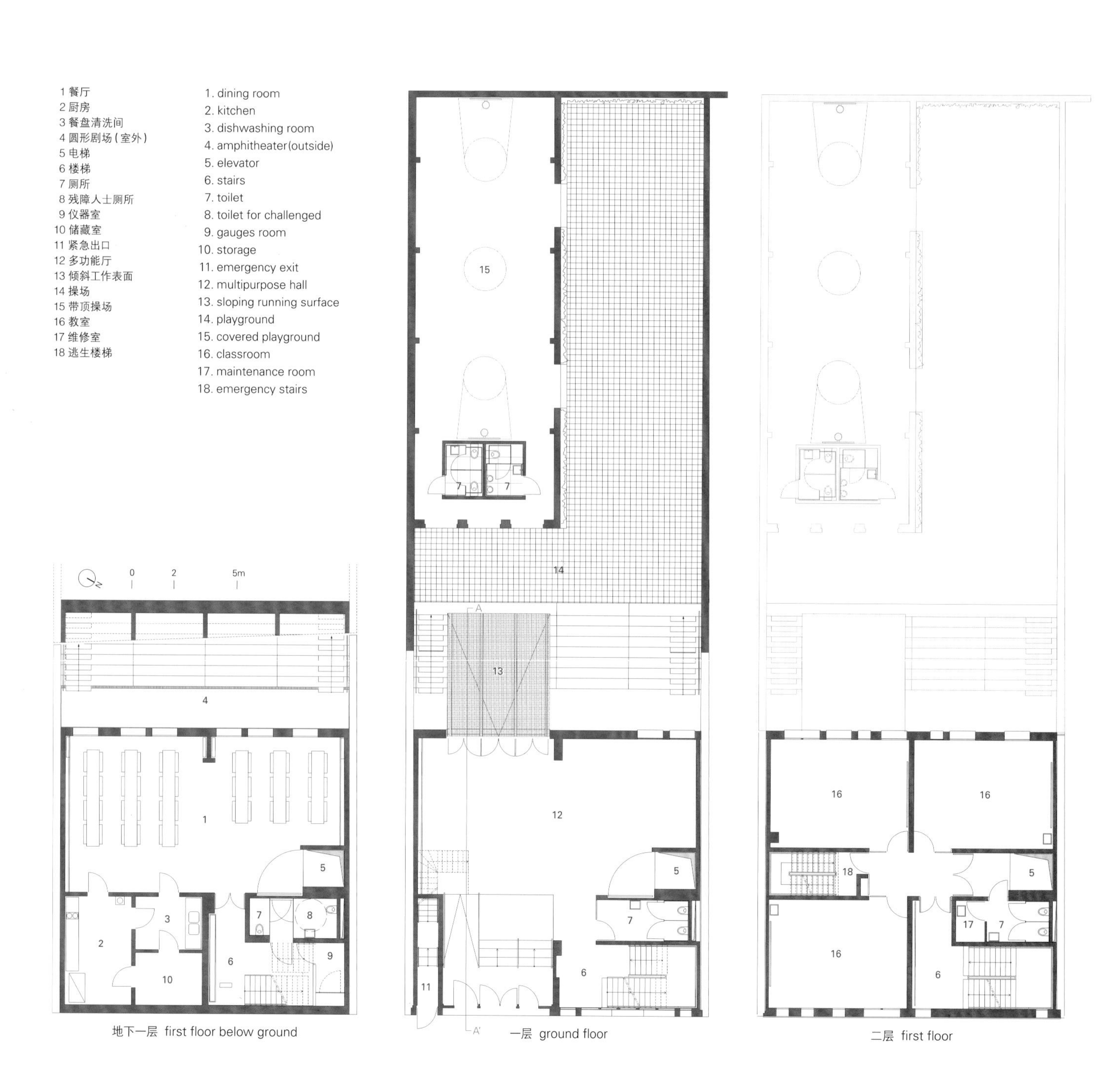

地下一层 first floor below ground

一层 ground floor

二层 first floor

1 餐厅
2 入口
3 多功能厅
4 倾斜工作表面
5 圆形剧场（室外）
6 逃生楼梯
7 学前班教室
8 小学教室

1. dining room
2. entrance
3. multipurpose hall
4. sloping running surface
5. amphitheater(outside)
6. emergency stairs
7. classrooms of preschool
8. classrooms of primary school

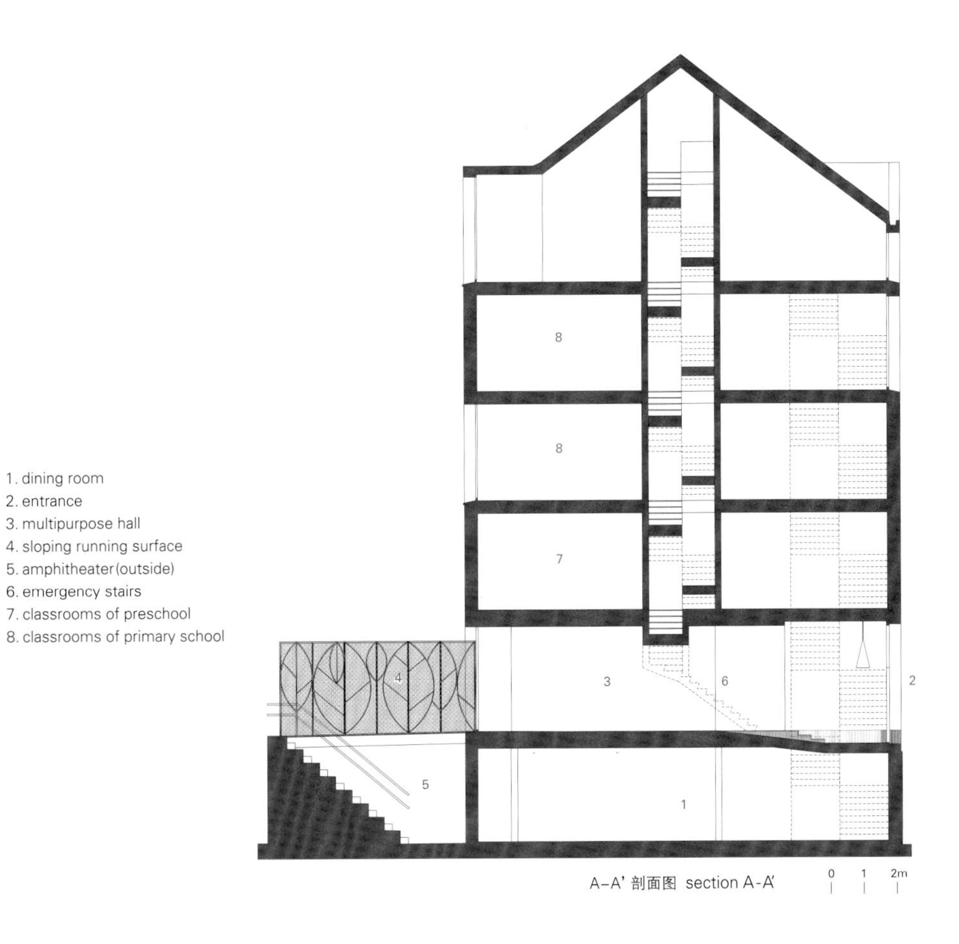

A–A' 剖面图 section A-A'

项目名称：Lagere School Sint-Gillis / 地点：Engelandstraat 49-51, 1060 Sint-Gillis / 建筑师：Bart Lens / 用地面积：1,035.80m² / 造价：EUR 2,272,294.35 / 竣工时间：2010
摄影师：©Philippe van Gelooven(courtesy of the architect)

后社区时代

After the Neighborho

与持续扩张、稀释和冷却的宇宙不同，第一世界的现代城市就算没有停止，也早已放缓了其离心进程，生产动力的深刻变革使发展的驱动力也已经逐渐停止。

只要离心扩张仍在发生，城市环境就不可避免地要经历以下时期：根据不断变化的需求在有限的区域内重新安置、新功能的出现和土地价值的增长或下降。

尽管如此，以前的发展进程还是留下了一点不连续性和城市退化的异常现象，反思扩张模式，现在则呈现出重要的拓扑价值；它们不再意味着放弃空隙，而是可以创造共同地标或自我表现的真正的场所和机会。这是新的入侵式建筑和综合城市结构之间新关系的起点。如果以前城市建设的主要规则是同系化和模仿，那么现在，集体努力和社区的本义已经消失，个体差异的渴望开始浮现，这些都标志着相异的语义基体，因而显得更加意义重大。

冒着以偏概全的风险，我们可以说，在这些环境下的每个新项目都可以被简化为两种倾向：影响环境，或者正在受到环境的影响。而且这两种倾向有可以共存的机会。因而，其中一些像沉默的观察者或表面装饰一样被受限地安置于现有的环境中。

如果说新建筑对环境漠不关心，那是不正确的。它们融入社区的态度正在发生变化，从一体化和隶属意识，转变为通过使用同一时代的词汇作为语言表达的契机，来捕捉潜在的场所精神。

这些例子可作为对形象化的社会、倡导者的极端研究以及特立独行方式的清晰表达。尽管如此，这些例子表达出了“在那里”的强烈愿望，使社区恢复活力，否则将逐渐失去生成它们的社会力量。

Unlike the universe, which continues to expand, thinning and cooling, modern cities of the first world have long since slowed, if not stopped, their centrifugal process, having halted progressively the driving force of growth as of profound transformations of production dynamics.

As long as centrifugal expansion was in action, urban contexts lived through phases of resettlement of limited areas according to changing needs, the emergence of new functions, and the growth or fall of land value.

Nevertheless, the previous process of growth has left behind small discontinuities, anomalies in the urban entropy that, in rethinking the expansive model, now assume an important topological value; they no longer represent abandoned voids but rather genuine places, opportunities to create collective landmarks or individual statements themselves. This is the starting point of a new relationship between the new intruder buildings and the consolidated urban fabric. If previously the main rule of urban construction was homologation and imitation, now, that collective effort and the original meaning of the neighborhood have been extinguished, a desire for individual distinction emerges, marking the anonymous matrix by an alien semantics, which thus makes it more significant.

At the risk of generalizing, we can say that each new entry in these contexts can be reduced to two attitudes:—affecting the context or being affected by it—and there is a chance that both attitudes can coexist. Some of them, then, are limited to settle as silent observers, encrustations on the existing environment.

If it is not correct to say that new buildings are indifferent to the context, the attitude with which they fit into the neighborhood is changing, passing from integration and the awareness of belonging to it to the will to capture the latent genius loci as an opportunity for linguistic expression, through the use of a contemporary vocabulary.

As clear expressions of the society of the image, of an extreme research of protagonism, and of distinction from the mass; nevertheless, these examples speak of a strong desire to "be there", revitalizing neighborhoods that otherwise would gradually lose the social forces that generated them.

od's Days

位置、环境和社区曾经是联系密切的概念。虽然我们可以认为"位置为主要的关联和环境，而不是标量或空间"[1]，但是依然不能低估自然环境的意义，因为自然环境是有助于位置和社区形成的主要因素之一。

古代的社区在社会和文化方面是简朴的，因为居住和构成社会的人是简单而相似的，其增长的特点是缓慢的聚集和分层，因而有助于形成总体形式上的连贯性。

由于强烈的归属和认同意识一直是特定社会识别为社区的独特要素，所以很显然，过去建筑环境中的构造物的主要标志是同系化，并共享类型化和形态上的城市模式。

如果这一事实对稳定的前工业城市更是如此，甚至在伴随着工业生产体系确定的不寻常的扩张过程中也是这样，那么作为城市被认定为更多或更少扩展社区的要素之一，类型学和形态同质化的标准仍然会保持很长一段时间。

然而，在扩张阶段，对劳动力不断增加的需求给城市带来了巨大的人流量，很难确定居民区的小社区是如何在建立空间参考模式方面起到积极作用的。

与此同时，外来人口多快才能建立文化和社会归属感，由土地出租的进程来决定，也取决于城市的环境。阿尔让·阿帕杜莱的考量仍在继续："主要的困境在于社区既是大环境，同时也需要并产生大环境。"[2]

工业和后工业社区代表了一个更复杂的环境，"其中环境中隐含了其他环境，以使每个环境都隐含在全局环境的网络中，这与在文本中暗示其他的文本并最终暗含所有的文本并不相同。"[3]

基于上述理由，正如在宇宙膨胀的早期，物质和能量熵分布的微小差异遗留下异常和不对称现象；与其相似，迅速扩大的工业城市也留下了一些未使用的空间，在势不可挡的发展推动下被及时填充、重建和替代。

Locality, context, and neighborhood used to be strictly connected concepts. Although we can consider "locality as primarily relational and contextual rather than as scalar or spatial,"[1] it is not possible to undervalue the relevance of the physical environment as one of the primary elements that contribute to the formation of localities and of neighborhoods.
The ancient neighborhood was socially and culturally simple, as the people who lived and constituted it were simple and homogeneous, and its growth was characterized by slow aggregation and stratification that favored a total formal coherence.
Since a strong sense of belonging and identification has been a distinctive part of that specific community recognizable as a neighborhood, it is clear that in the past the construction of the built environment was marked by homologation and by the sharing of typological and morphological urban models.
If this fact was particularly true for the static pre-industrial cities, even during the extraordinary expansion that accompanied the affirmation of the industrial production system, it remained for a long time the criterion of typological and morphological homogeneity among parts of the city identified as more or less extended neighborhoods.
However, in that expansive phase, during which the increasing demand for labor attracted to the city a huge human flow, it is difficult to establish in which measure the small communities of neighborhoods have been an active part in establishing patterns of spatial reference.
At the same time, how quickly the immigrant populations built a sense of cultural and social belonging, also according to an urban environment, has already been determined by the processes of land rent. By continuing with the considerations of Arjun Appadurai, "the central dilemma is that neighborhoods both are contexts and at the same time require and produce contexts."[2]
The industrial and post-industrial neighborhood represents a more complex environment "in which contexts imply other contexts, so that each context implies a global network of contexts, is different from the sense in which texts imply other texts, and eventually all texts".[3]
For the above reasons, as in the earliest times of the expand-

Drawing©Vo Trong Nghia Architects

如今，对于生产体系的全球反思和有形商品制造从高科技、高福利国家转移到劳动力成本低、社会安全体系较不完善的国家，已造成了新旧大都市混乱和毫无计划的发展。同时，第一世界国家的城市对新住宅的传统需求下降，从而对疯狂的建筑购买行为和对建筑功能的要求持续不断的增加产生了显著的冷却作用。社区本身已经失去了它的历史重要性和统一的意义及其与环境的关系。

经过一段时间的衰落，由于原来社区的社会组成与城市的中央部分及其形成的母体本身逐渐疏远，新的参与者都对搬迁这些具有丰富潜在环境内容的城市地区产生了兴趣。在上述提到的创建环境的互惠机制中，在某些情况下，引进的新居民和新特点是新环境的创始核心。在其他情况下，它们被整合到一个强大的现有框架中。还有一些更复杂的情况，那就是新建筑与传统建筑找到了一个共存的方法。

在新世界或旧世界、静止或增长的经济体中，拥有几百年历史的老社区或那些最近新形成的社区代表了在各自情况下对新建筑和新移民而言丰富的文化内涵媒介，取决于复杂的分层进程，这与基本处于危机中的20世纪理性主义城市规划的观点明显不同。

忽略微不足道的社区整合完工案例，通过在构造形式上模仿却又缺乏原有特色和动力的建筑物，我们正在目睹这样一个有趣的现象：在一个充满活力的项目中，新的参与者选择成为这些环境的一部分或者从环境中生存下来的部分。

我们正在见证一种强烈而明确的当代建筑语言对公共和私人领域的过多介入，可以确知这些环境下新生命力的可能性，尽管它们就一些社区的持续特征留下了一些质疑，这些社区存在的目的在于研究同类人群组织的传统意义。

因此，新的干预项目似乎减弱了前文提到的地域、环境和社区在概念上的密切关系，甚至到了这样的地步：在一些情况下，它被确定为

ing universe, small discrepancies in the entropic distribution of matter and energy left behind anomalies and asymmetries; in a similar way, the rapidly expanding city of industry left behind it some unused spaces, which were promptly filled, reconstructed, and replaced under the thrust of unstoppable development.

Today, the global rethinking of the system of production and the relocation of manufacturing of material goods from countries with advanced technology and expensive welfare to those offering manpower at low cost and with fewer security controls have produced chaotic and unplanned growth of the old and new megalopolis developed through production processes. At the same time, the traditional demand for new homes in the cities of the First World has fallen, thus producing a significant cooling of expansion and of the continuous and frantic turnover of buildings and functions. The neighborhood itself has lost its historical and consolidated meaning and its relationship with context.

After a period of decline, due to the gradual estrangement of the social components of the original neighborhoods from the central parts of the city, from the matrix itself of their formation, new players have developed an interest in relocating in those urban areas rich in potential contextual contents. In the aforementioned mutual mechanism in creating the context, the new inhabitants and the new features introduced are, in some cases, the founding nucleus of a new environment. In other cases, they are integrated into a strong existing framework. In still other more complex situations, the new and the traditional find a way to coexist.

In the Old or New World, in stationary or growing economies, centuries-old neighborhoods or those of recent and very recent formation represent in each case a rich culture medium for new insertions and new immigrants according to a complex process of layering that is clearly different from the idea, basically in crisis, of the planned city of twentieth-century rationalism.

Neglecting the cases of trivial completion of consolidated neighborhoods through the construction of buildings that mimic forms now devoid of their original character and motivation, we are witnessing the interesting phenomenon of a highly motivated entry of new players who choose to be part of those contexts or of what survives of those contexts.

We are witnessing a plethora of such public and private insertions that, by a strong and unequivocal contemporary

复杂的共存形式，各层之间不一定是彼此连通的。

然而，随着这个复杂问题的普遍限制，我们可以想象，这些建筑物环境中的干预项目能产生各种不同的效果。在一些情况下，强大的剩余价值的持久性会影响新干预项目的类型学、形态学和组成特征；在其他情况下，这种新加入的元素代表了已经部分或完全失去的环境再生的原核。此外，像查尔斯·西蒙兹设计的微型住宅，嵌入大都市的一个偏僻角落[4]；一些新的建筑，如外国企业，占据了实际上不是一部分的空间；还有一些聚居地，只是有人居住在那里而并没有在所有方面都成为构成要素。

无论如何，这些公共和私人项目的介入态度，就算在传统意义上不算社区，也是环境的一部分，从来都不是简单的整合和寻找归属感，而是抓住机会在语言上表示肯定，用于当代的权利诉求，而不是诉诸虔敬或是各种地方语言。

在这个意义上，这些介入设计清楚地表达了形象化的社会，以及对获得个体肯定和主人公身份的强烈愿望，但同时，在对城市史如此丰富的定居地点进行选择时，它们通过新形式回答了对"在那里"和恢复进程的需要，否则这一进程将不可避免地遭到破坏或者充其量变得像博物馆。

即使有必要区分，并且没有硬性分隔，以下实例也代表了各种不同的途径。概括来讲，前三个可作为干扰项目影响环境的例子，后三个作为对环境加以考虑并融入环境的建筑实例。

一类建筑希望通过自身强大的存在来影响其环境，由Artech建筑师KwanSeok Kim所设计的Artech住宅正属于这一类，位于首尔大都市的连续网格中的侵入式生混凝土体块。

虽然它时尚的倾斜而带角度的表面使其能够被看作介入环境中的一个入侵者，但是它与老旧混乱的江北区结构的关系也不是那么陌生，

language, can confirm the possibility of a new vitality of those contexts, although they leave some doubt regarding the continuing nature of the neighborhood intended in the traditional meaning of homogeneous anthropological entity.

As a result, the new interventions seem to weaken the close relationship among the concepts of locality, context, and neighborhood that were mentioned above, to the point that it is determined, in some cases, to be a form of complex co-existence among layers not necessarily communicating with each other.

Within the limitations of the generalization of such a complex matter, however, we can imagine that interference in the context of these buildings can produce widely different effects. In some cases, the persistence of strong residual values affects the typological, morphological, and compositional characters of the new intervention; in other cases the intruder element represents the germinal nucleus of a regeneration of contexts now partially or completely lost. Moreover, like the micro dwellings by Charles Simonds encrusted in the remote corners of the metropolis[4], some of the new buildings occupy, as foreign entities, spaces of which they are not actually part, localities that they inhabit only physically without being constituting elements in all respects.

In all cases, the attitude by which these insertions, both public and private are part of the context, if not of the neighborhood in a traditional sense, is never of simple integration and belonging, but seizes the opportunity for a linguistic affirmation, for a claim of right to the contemporariness, without resorting to pietism or vernacular languages.

In this sense, these insertions represent the clearest expression of the society of the image, a strong desire for individual affirmation and protagonism, but at the same time, in choosing to settle in localities so rich in urban history, they answer to a need for "being there" and for resuming, by means of new forms, a process that otherwise would inevitably be broken or, at most, museumificated.

Even with the necessary distinctions, and without rigid compartmentalization, the following examples represent these different approaches. Schematically, the first three may be taken as examples of intrusion affecting the environment, the last three as buildings that take it into account and are incorporated into it.

To the category of buildings that want to influence the context with their strong presence belongs the Artech House, an

同时对这个缺乏系统性的城市体系表达出了需要给予更多感知和确定模式的诉求。然而，其中精确的构造清楚地表达出不完全依附于环境的关系。连接大楼到地面上两个楼层的变形虫状空间看起来像一个拒绝与当地几何结构相连接的防御构造。通过舷窗向外部开放的决定突出了在附近"浮动"的感觉。只有高耸的上部楼层恢复了与空气和光线的关系，在体积和开口方面继续利用了线性和垂直关系，更好地适应周围的环境。

荷兰的格罗宁根体育大厦是一个包括两个运动场馆的综合体。该项目的公益性质是用以确定新城市价值的生产者在社区中的作用。尽管作者的本意并不想和周边环境偏离太远，但是其大幅度起伏的立面和水平外形，与相邻建筑物的垂直走势相比，再考虑到纵向纹理的道路，不可避免地形成了一个独特之处。对于举办体育活动而言，并不需要在大面积的砖墙上设置开口，这一要求导致了另一个突变性，意味着对当地的形象产生了极大的改变。

与此相反，很难确定由Vo Trong Nghia建筑师事务所设计的、位于越南胡志明市谭平区的树之家的闪烁构想，是将其本身施加于城市结构之上，还是中肯地接受城市结构，并表现为一种超规模的绿色装饰。如果作者的意图是通过其设计的介入，给谭平区的过度污染区带来越南热带森林的绿色空间，那么看起来，为了实现让住宅区成为巨大的"阳台花盆"这一目的，采用的理念是调节大树与典型的常见元素的自然特性。

另一方面，一家由Zemel+建筑师事务所设计的编辑工作室——Luis Anhaia工作室附近的入口，则充分尊重了南美大都市圣保罗市维拉马达莱纳区的纵向纹理。

虽然由当代直线语言定义，但该建筑也注意到了当地的自然状况，特征在于窄而深的切割地块，并展示出与附近地区的以往历史无

aggressive block of raw concrete in the endless grid of the Seoul megalopolis, by KwanSeok Kim of Artech Architects. Though its trendy oblique and angular surfaces make it possible to think of it as an intruder inserted in the environment, its relationship with the old and chaotic fabric of the GangBuk area is not so alien and expresses the need to impose more aware and determined patterns on that unsystematic and entropic urban system. However, the precise composition of the volumes clearly expresses a relation of not total adherence to the context. The amoebous volume that links the building to the ground for the first two floors looks like a defensive structure that refuses to connect to the local geometry. The decision to open up to the outside through portholes accentuates this feeling of "floating" in the neighborhood. Only the upper floors, towering and regaining the relationship with air and light, resume, in volumes and openings, a linearity and squareness that better suit the surroundings.

In the case of the Groningen Sport Block in the Netherlands, a complex that includes two sports halls, it is the public nature of the project to determine the role of generator of new urban values in the neighborhood. Despite the authors' intention not to stray too far from the general surroundings, the scale of the large undulations of the facade and its horizontal development compared to the vertical trend of the adjacent buildings inevitably create a singular point with respect to the ordinate texture of the road. The need to insert no opening in the massive brick wall – openings are undesirable in relation to the hosted sporting activities – creates another discontinuity that implies a strong revision of the image of the locality.

On the contrary, it is difficult to determine whether the winking idea of Vo Trong Nghia Architects for the House for Trees in Tan Binh District, Ho Chi Minh City, Vietnam, imposes itself on the urban fabric or accepts it to the point of representing an out-of-scale green decoration of it. If the intention of the authors was, by their admission, to bring to the over-polluted area of Tan Binh District the green space of Vietnamese tropical forests, it seems that the concept adopted for this purpose – residential volumes treated as huge "balcony pots" – is imagined to mediate the naturalness of the large trees with typically familiar elements.

The entry into the neighborhood of the Luis Anhaia Studio, an editorial atelier by Zemel+Arquitectos, on the other hand, is fully respectful of the ordinate texture of the district of Vila Madalena in the South American megalopolis of São Paulo.

1. Arjun Appadurai, *Modernity at Large: Cultural Dimensions of Globalization*, University of Minnesota Press, 1996.
2. ibid.
3. ibid.
4. http://www.charles-simonds.com/dwellings.html

缝衔接的整合能力。与完全遵循周围纹理规则相匹配的，是对内部空间的自由阐释。其内部空间被设定为开放的空间，超越了因为狭长的传统地段而造成的功能上的限制。

离开圣保罗这种特大城市，来到比拉塞卡小镇，这里几乎可被视为一个独特的街区，VV Arquitectura建筑师事务所在拉瓦尔德拉玛尔酒店的设计上也面临着相似的情景，即由一系列狭长地段构成的哥特式纹理的相似主题。这种结构产生于社会充满不安全感的时代，促进了社会和地理环境的密集度，并且很容易从外面进行防卫，但是很难适应现代生活的需求。正如在Luis Anhaia工作室里，即便以当代眼光做出修改后并与多个地段进行了整合，内部空间的开放式布局也与大幅度接纳周围空间的规则相匹配，这样就使得大户型高层客房的建造适用于迎接宾客。这个功能引入了一种充满挑战和永远新鲜的存在，恢复了历史脉络。

与上述相似，由SMF建筑师事务所设计的布宜诺斯艾利斯大地测量学历史档案馆，嵌入了这座典型殖民地新世界城市的辽阔而坚硬的土地中，受到了狭长地段的限制和影响。然而，与上述相反，在这里仅限于对于场地布局以及体量高度大大超过周围较低的艺术装饰建筑这一情况的接纳。这样一来，本项目的体量上升并超过了别的建筑，作为一个显著的地标影响它们，同时也受到城市布局的影响，展现出模糊的关系，这种关系建立在由缓慢的历史进程所形成的区域基础上。

对上述各个项目的分析，清楚地展示了层次丰富、组织分明的城市环境的复杂性，也体现了为了保守的保护观念而能够打断形成过程这一想法的空白无力。

Although defined by a linear contemporary language, the building takes note of the nature of the place, characterized by the narrow and deep cut of the land parcels, demonstrating the capability to integrate with continuity and without trauma in the neighborhood's previous history. The full compliance with the rules of the surrounding texture is matched by a liberal interpretation of the interior space, which is organized as an open space, going beyond the functional limits imposed by the acceptance of the traditional long and narrow lot.

Leaving the megalopolis of Sao Paulo for the small town of Vila Seca, recognizable almost as a unique neighborhood, VV Arquitectura also faces, in the Raval de la Mar Hotel, a similar theme of Gothic texture consisting of a series of long and narrow lots. This texture, born in times of great social insecurity, suggested a social and geographical environment dense and easily defended from the outside, but poorly compatible with the needs of modern life. As in Luis Anhaia Studio, the substantial acceptance of the rules of the surrounding space, even revised with contemporary eyes and the combining of several lots, is matched by an open organization of the interior space, which enables the creation of large and airy rooms appropriate to the receptive function. This function introduces a challenging and ever-new presence, reviving the historical context.

Inserted in the vast and rigid centuriatio of a typical city of the colonial New World, the project by SMF Arquitectos for the Historical Archive of Geodesia Direction of the Province of Buenos Aires is affected, like the ones above, by the constraints of a long and narrow lot. Contrary to the above, however, the acceptance is here limited to the layout of the site, and the volume develops to a height that is relatively off scale from the surrounding lower art deco buildings. In this way, the volume rises over the others, affecting them as a significant landmark while being affected by the urban grid, demonstrating the ambiguity of the relationships that are established in areas formed by the slow historical process.

The analysis of the mentioned projects clearly shows the articulated complexity of the stratified urban environments, and the vacuity of the idea of being able to interrupt the process of formation on behalf of a conservative idea of protection.

Aldo Vanini

Drawing©Vo Trong Nghia Architects

Artech住宅

Artech Architects

산공영주차장

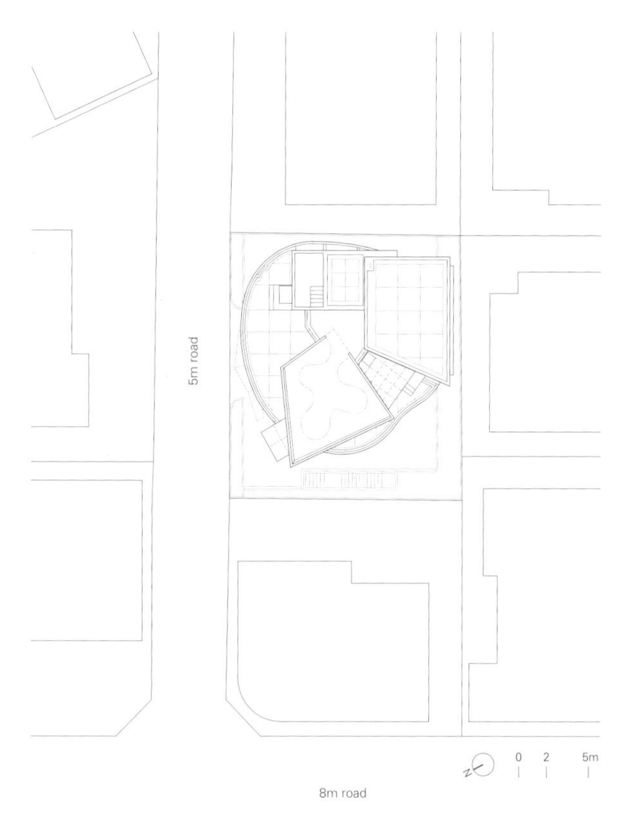

汉江把韩国首尔市分成两个部分：江南区和江北区。与新兴的江南区相比，北部的江北区保持着传统风貌——旧日的王宫、传统的瓦屋、古朴的建筑、古老的市集。有着500年历史的汉城古城位于中心地带，周围是杂乱的旧城区，中间耸立着新建的高楼大厦。

Artech住宅位于日本殖民政府20世纪30年代修建的马路边，这里与江北区其他中心住宅区不太一样。几十年的无序发展破坏了东北地区的山丘，现在被划定为重建区。

在这种情况下，能否将现代技术和美学设计融入建筑中，使人们能在那里安居乐业？该项目要在普通的方形砖瓦建筑旁边的一座与众不同的三层办公楼顶设计独立屋。独立屋的设计体现了自由，实现了东西方共有的独特主题。我认为现在首尔地面上的独立屋渐渐消失了，所以我想在办公楼顶部设计形式自由的独立屋。

进入房屋前的单行路，一楼是停车场，左边是Artech住宅的入口，右边是固定长椅和地下办公室的锥形天窗。二层和三层外观独特，形式自由的弧形外墙上装点着大大小小的圆形窗户。当夕阳射进窗户时，这些圆窗产生有趣的效果，周围杂乱的景物显得井井有条。大楼的点睛之笔是四楼的室内游泳池，游泳池也是办公楼上方独立屋的一楼。

泳池巨大的花格窗贯穿四到六楼，透过窗户，壮丽的江北自然公园——南山的景色尽收眼底。

六楼卧室对着小型的顶楼花园。家人可以在顶楼的野花花园里畅游，欣赏南山、江北城区风光和北部群山。群山、楼宇、碧空，全角度展现在眼前。

Artech House

Seoul, South Korea is divided into two large districts by the Han River: the southern part is called GangNam. In contrast with newly developed GangNam, GangBuk which is the north part has historical palaces and traditional houses with old buildings in the old downtown area, including the 500-year-old Hanseong Castle in the middle, while the periphery has old reckless development area in between newly built apartments.

Artech House is located on the road built by the Japanese colonial government in the 1930s, unlike the rest of central residential area of GangBuk. The neighboring hills of the northeast have been harmed by decades of reckless development and are now designated a redevelopment zone. In these circumstances would it be possible to incorporate modern technology and artistic design into a building where people can happily live and work? This project was designed to house a single-family home on the top of a distinctive three-floor office building next to the common square buildings of brick or tile. The design of single-family homes has realized freedom and unique themes in the East and West alike, but for me many single-family homes on the ground seem to be gradually disappearing in Seoul nowadays. So I wanted to design a free-form single-family home on the roof of the office building.

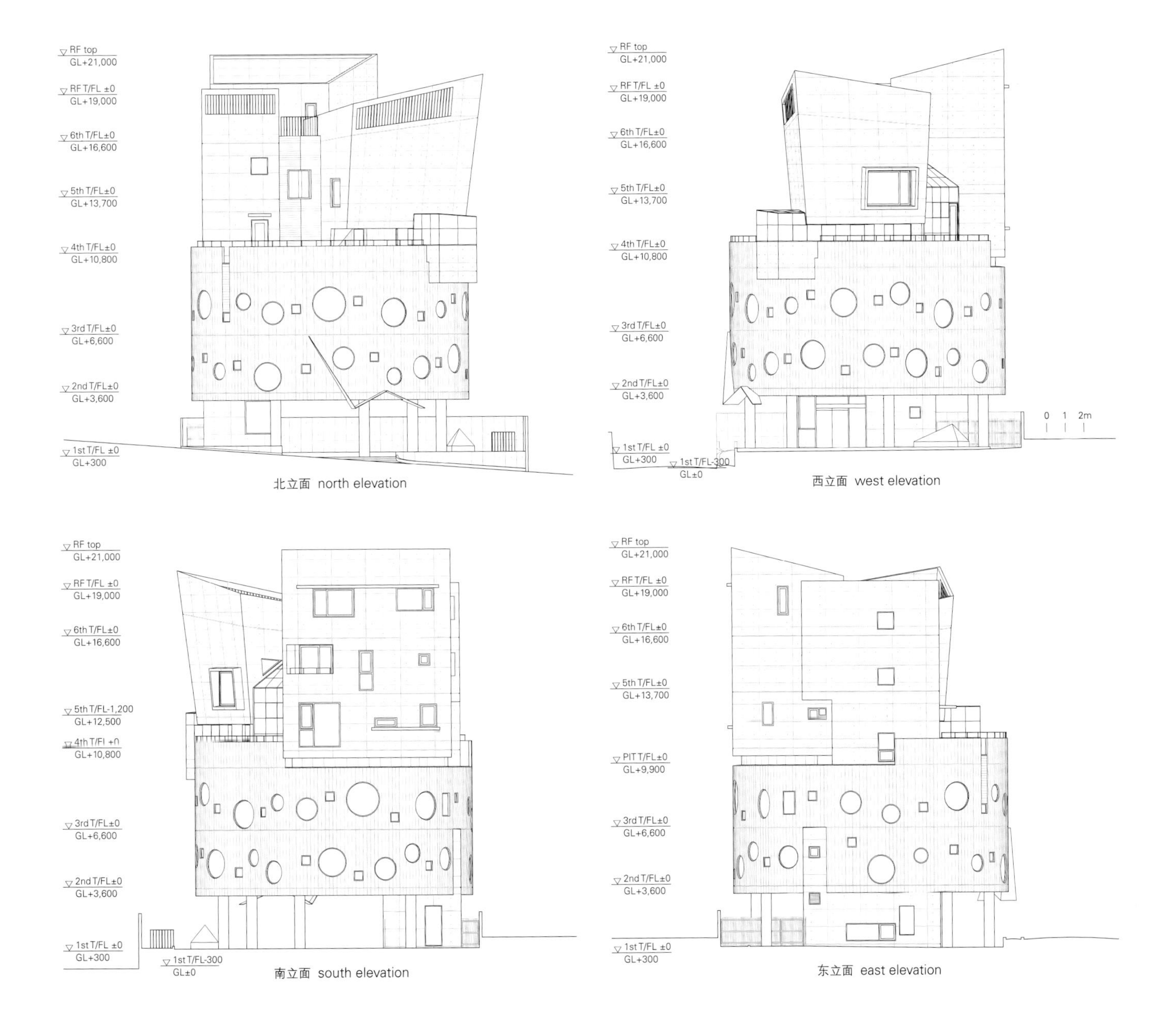

北立面 north elevation

西立面 west elevation

南立面 south elevation

东立面 east elevation

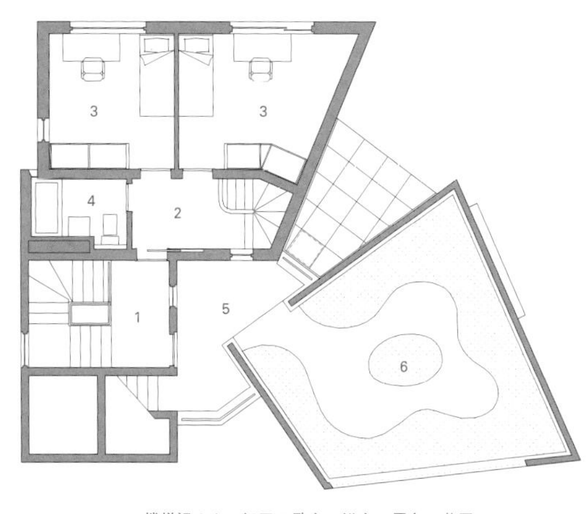

1 楼梯间 2 入口门厅 3 卧室 4 浴室 5 平台 6 花园
1. stair hall 2. entrance hall 3. bedroom 4. bathroom 5. deck 6. garden
六层 sixth floor

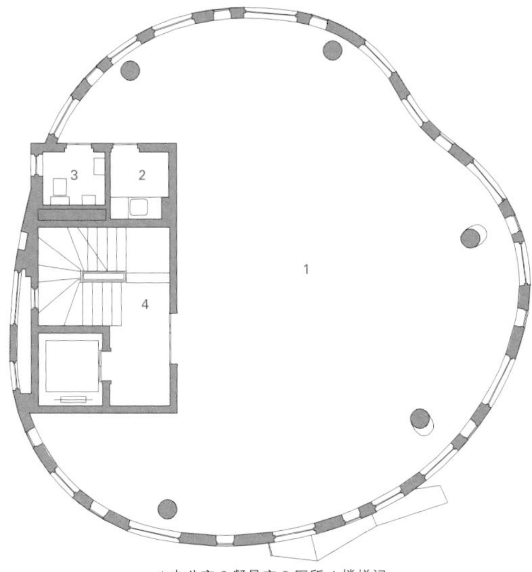

1 办公室 2 餐具室 3 厕所 4 楼梯间
1. office 2. pantry 3. toilet 4. stair hall
二层 second floor

1 楼梯间 2 入口门厅 3 主卧 4 更衣室 5 化妆室 6 起居室 7 书房 8 浴室 9 阳台
1. stair hall 2. entrance hall 3. master bedroom 4. dressing room
5. powder room 6. living room 7. study 8. bathroom 9. balcony
五层 fifth floor

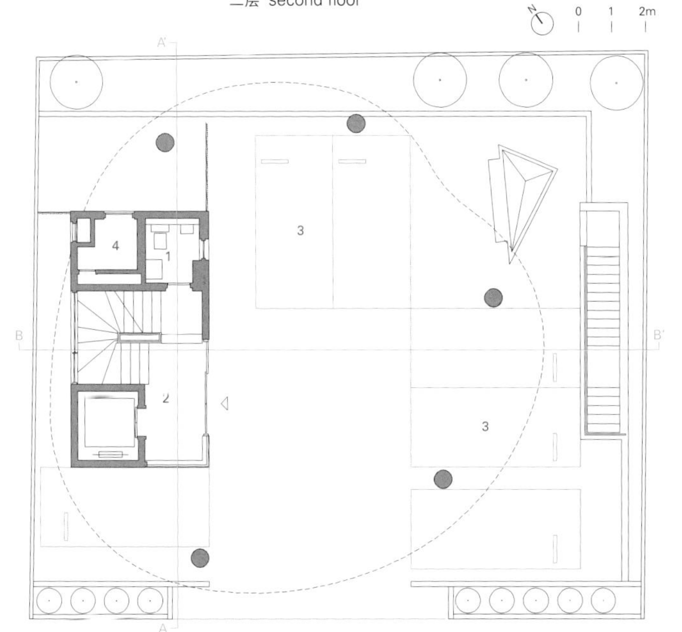

1 厕所 2 楼梯间 3 停车场 4 机械设备间
1. toilet 2. stair hall 3. parking space 4. mechanical room
一层 first floor

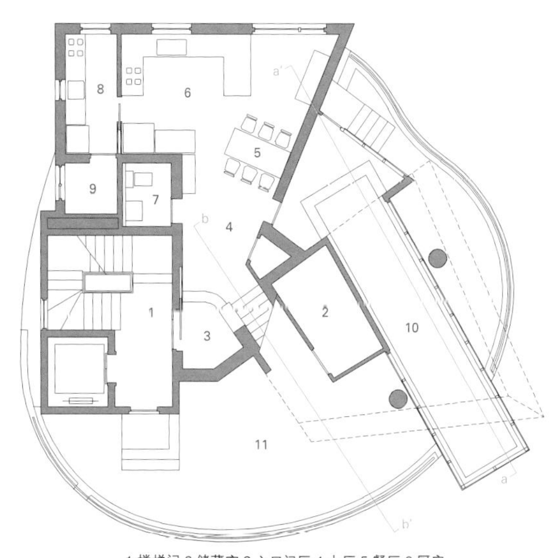

1 楼梯间 2 储藏室 3 入口门厅 4 大厅 5 餐厅 6 厨房
7 浴室 8 杂物间 9 锅炉房 10 游泳池 11 露台
1. stair hall 2. storage 3. entrance hall 4. hall 5. dining 6. kitchen
7. bathroom 8. utility 9. boiler 10. swimming pool 11. terrace
四层 fourth floor

1 办公室 2 餐具室 3 水箱 4 主卧 5 卧室 6 化妆室 7 厨房 8 浴室 9 露台 10 野花花园
1. office 2. pantry 3. water tank 4. master bedroom 5. bedroom
6. powder room 7. kitchen 8. bathroom 9. terrace 10. wild flower garden
A-A' 剖面图 section A-A'

1 办公室 2 会议室 3 停车场 4 储藏室 5 起居室 6 游泳池 7 露台 8 野花花园
1. office 2. meeting room 3. parking space 4. storage
5. living room 6. swimming pool 7. terrace 8. wild flower garden
B-B' 剖面图 section B-B'

46

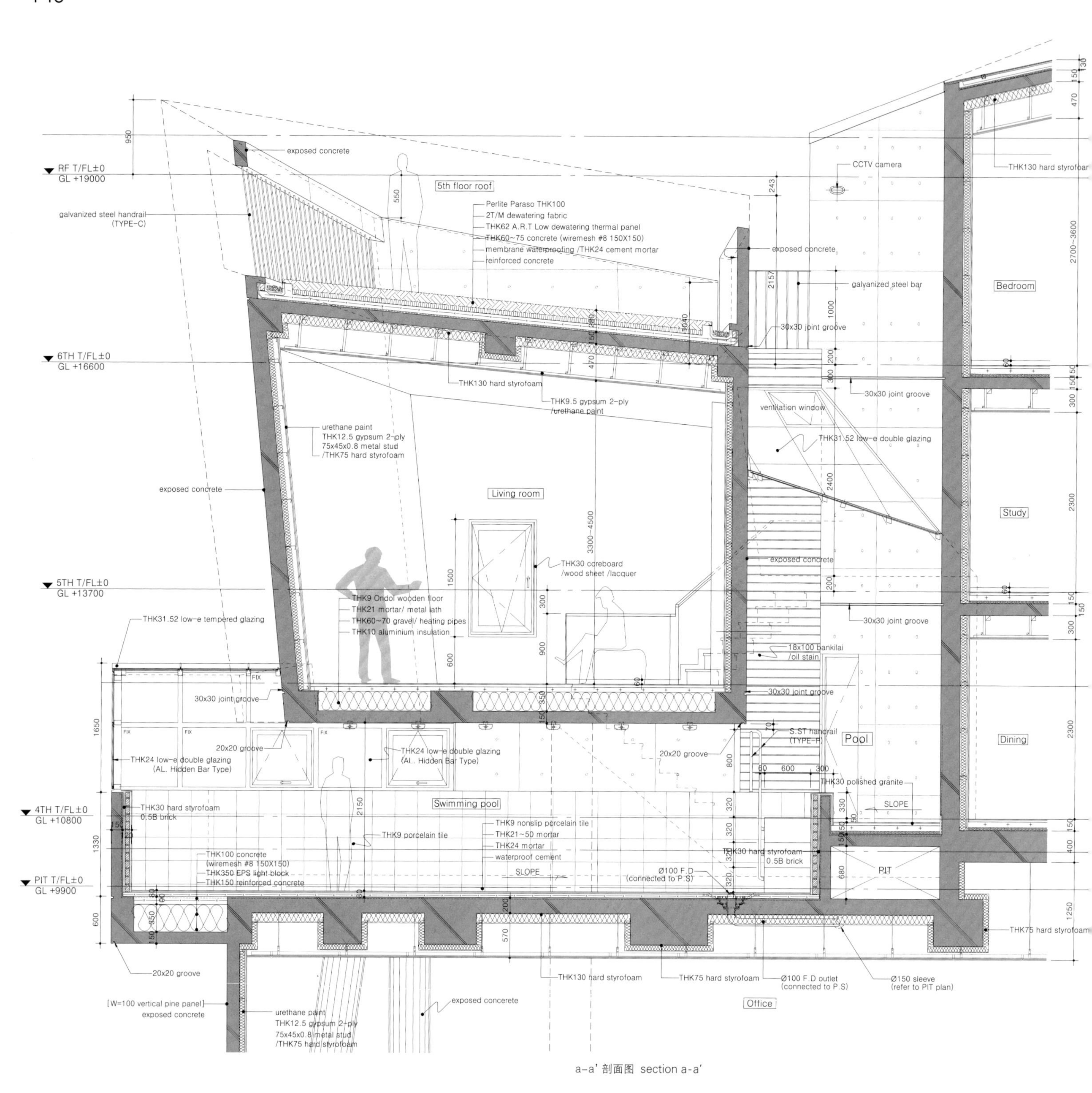

a-a' 剖面图 section a-a'

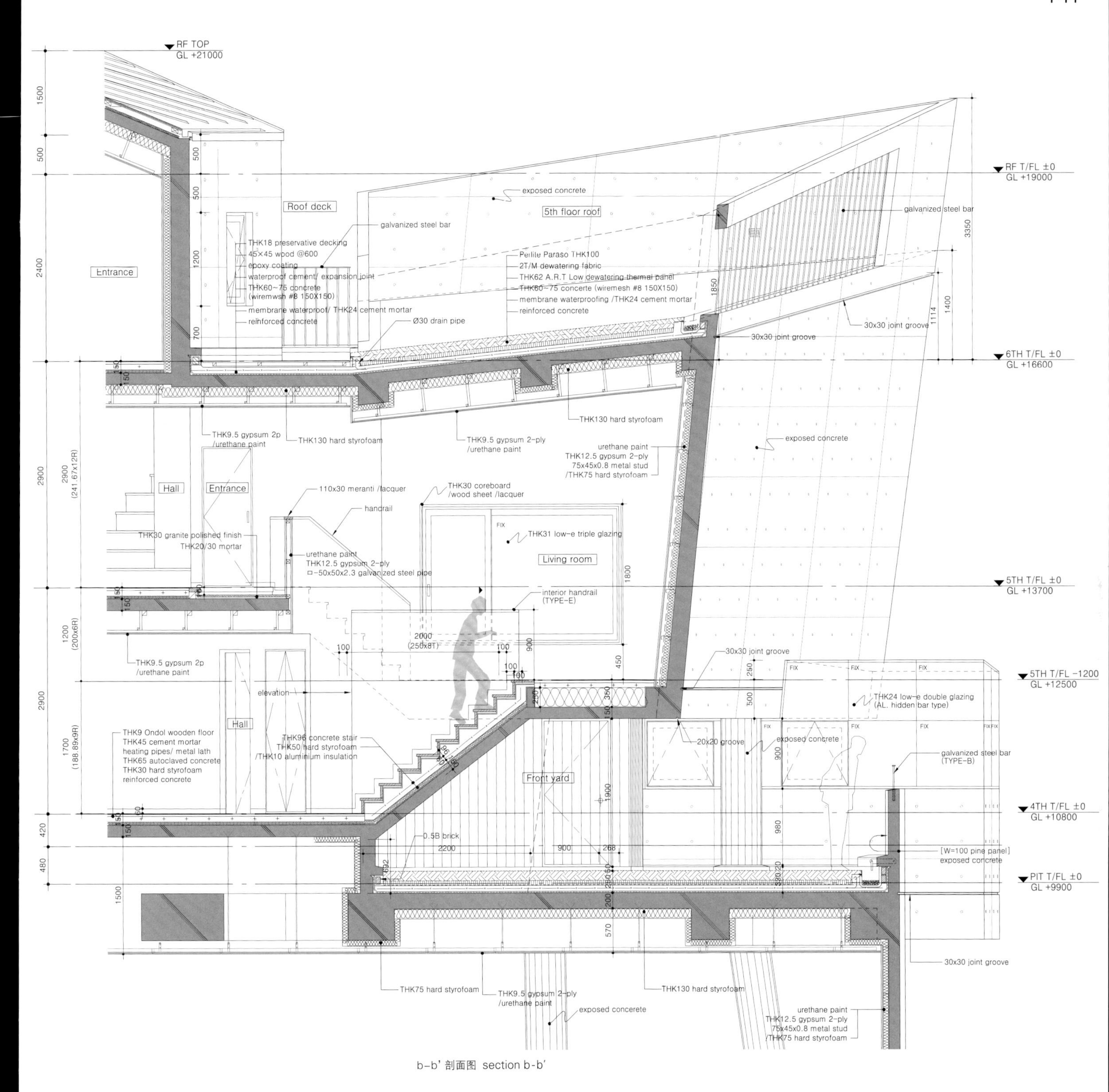

b-b' 剖面图 section b-b'

Upon entering the ground floor from the one-way road in front of the building, there is a parking lot. On the left side stands the entrance of Artech House. On the right side is a built-in bench with trigonal pyramidal skylights of the underground office. The second and third floors have a peculiar exterior characterized by a free-form curved surface with circular windows of various sizes. These windows bring order to the chaotic landscape around the building, while displaying an interesting effect when the evening sun passes through them from the southeast. The highlight of this building is an indoor swimming pool on the fourth floor which is part of the ground floor of the single-family house on top. Through the large lattice windows of the swimming pool, which connects concrete from the fourth to sixth floors, Nam Mountain, the natural park of GangBuk comes into sight in a splendid way. Finally on the sixth floor, bedrooms open onto a small rooftop garden. On the tilted roof a wildflower garden was created for the family to explore while enjoying the view of Nam Mountain, the GangBuk cityscape and the northern mountain. It offers a beautiful panoramic view of the mountains, surrounding buildings and the refreshing sky above the garden.

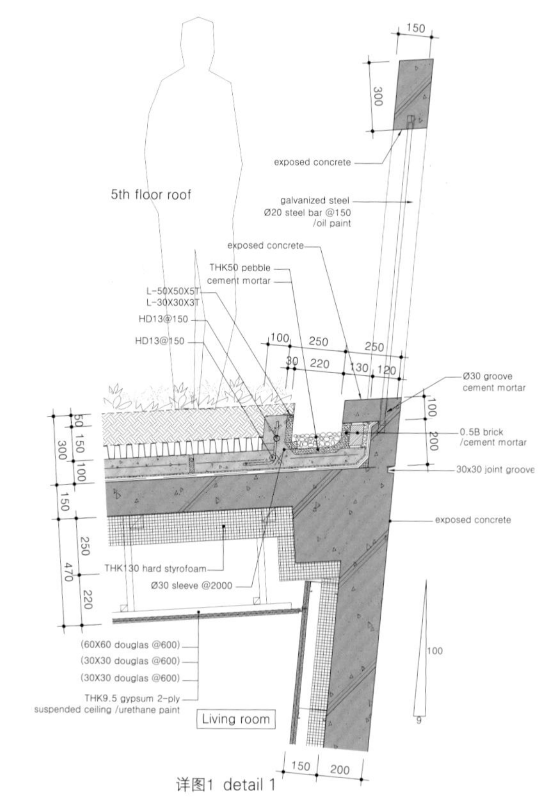

详图1 detail 1

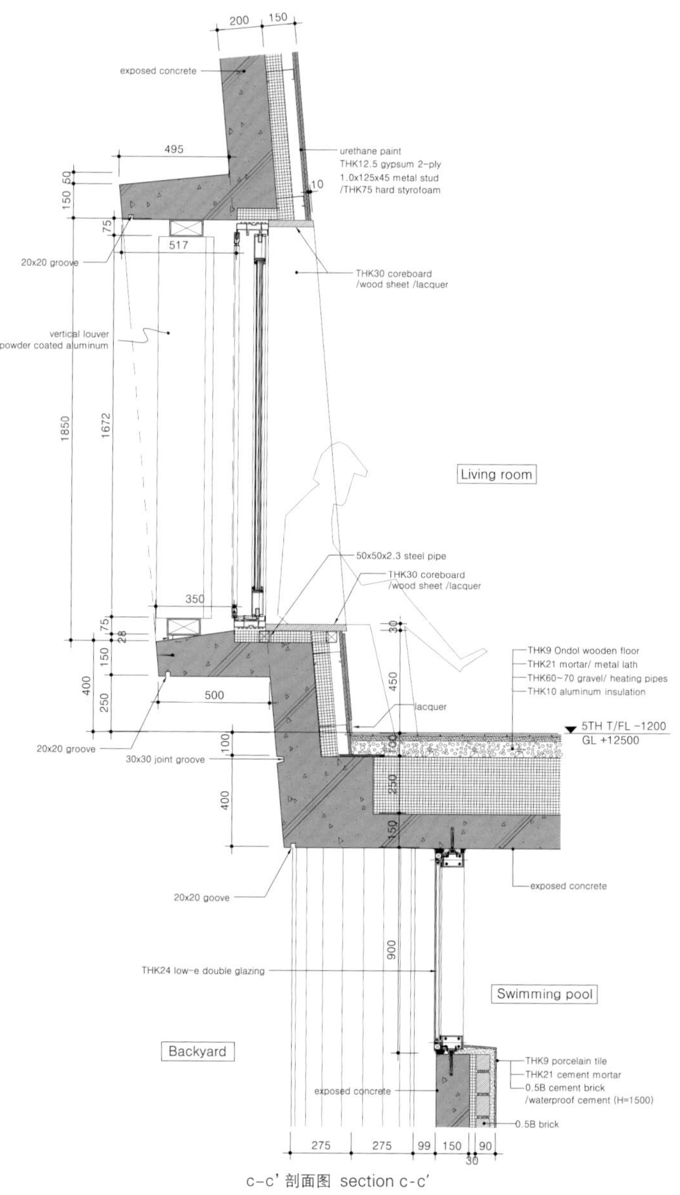

c-c' 剖面图 section c-c'

项目名称：Artech House / 地点：46, Dasan-ro 22-gil, Jung-gu, Seoul, Korea
建筑师：KwanSeok Kim _ Artech Architects
设计合伙人：SangHo Moon, YoungHo Kim, JeongBae Choi
展示：JinA Gang, JeongBae Choi
结构设计：SeongJae Lee _ Dan Stucture / 施工主管：SuYong Won
结构：reinforced concrete
用地面积：298m² / 建筑面积：168.11m² / 总楼面面积：775.61m²
高度：21m / 设计时间：2013 / 竣工时间：2014
摄影师：©JongOh Kim

格罗宁根体育大厦

Architectenbureau Marlies Rohmer

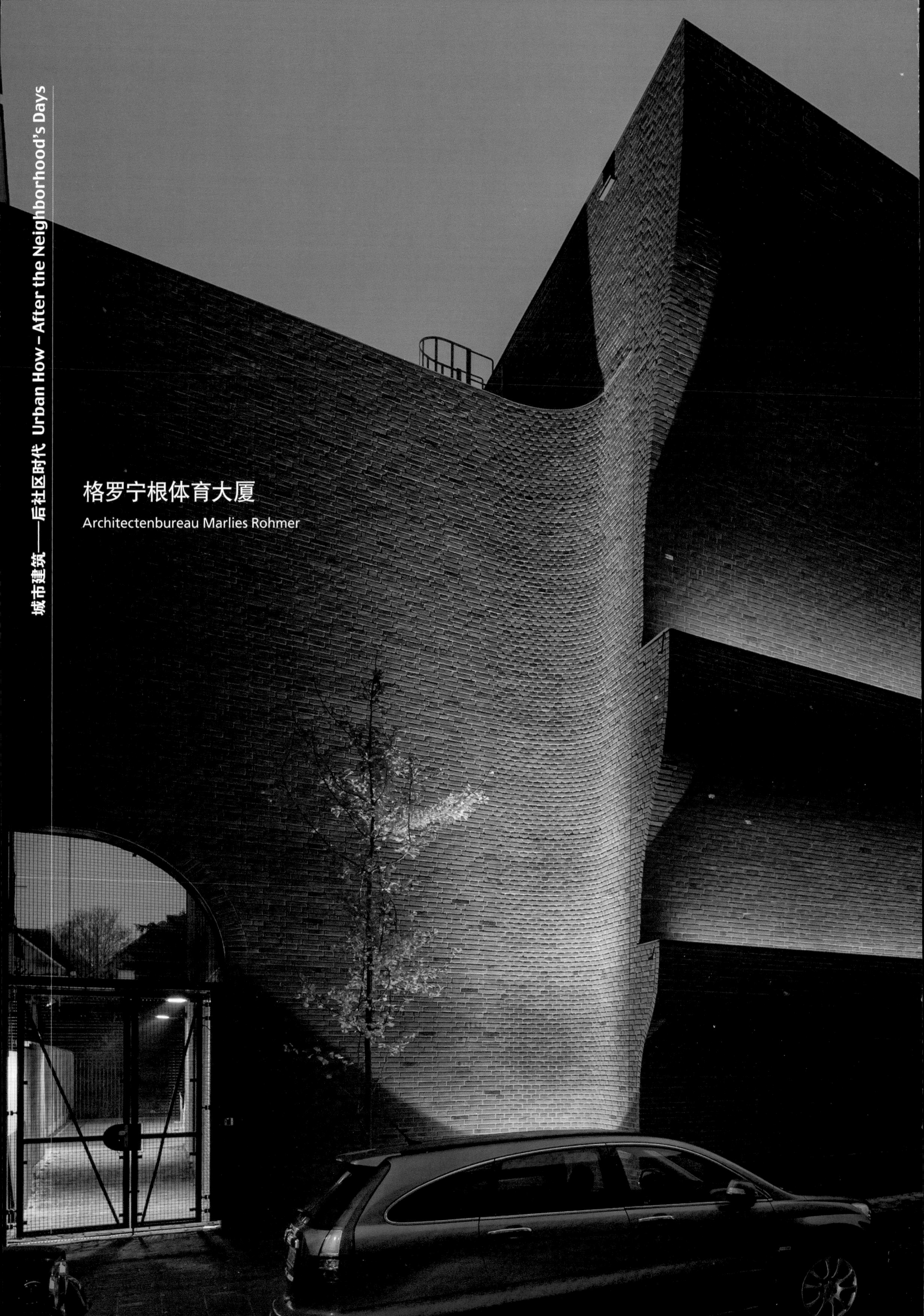

格罗宁根的小提琴街是一条铺着细密砖石的城市街道，如今这条街道上坐落着一座相对完全独立的综合设施，它包含两座体育中心大厦。分区规划和功能规格中潜在的冲突最终形成了一座多层建筑的设计方案。

事实证明，两座体育中心大厦并行在一条直线上的设计优势多多。这样的设计使得大厦对街景产生相对适度的影响，也使得在远离周围房产的场地上建造一个建筑体量成为可能，因此没有给人们留下一种庞然大物的印象。建筑两侧栽种着树木的两个小广场凸显了这座城市综合设施的植入。由建筑的拱形大门可以进入设置有自行车棚的后方区域。学生们可以在进入体育中心之前聚集于这些空间。体育大厦因此在没有任何特殊设计方案的情况下，成功地融入这条街道细致的城市纹理当中。

一般而言，体育中心都是不透明的箱状结构。设计师通常会将直射的阳光看作是体育中心设计的一种不利条件，因为直射阳光很刺眼，容易形成视觉干扰。而另一方面，在没有阳光的情况下运动是不利于健康的，因为它不符合人的生物节律。因此，大厦的设计采用波状的砌砖立面，使间接光线能够照射到每一个体育馆。

这个坚固而又不那么透明的体育大厦的体量看上去似乎想与现有街道的建筑风格形成反差。然而所应用的建筑材料也不应与周围一般环境所用的材料相差太多：红色"格罗宁根"砌砖形成的波状外表。当体育中心在夜晚投入使用时，因为使用了间接光线，建筑的立面炫彩夺目。带有双层高度的宽敞透明的楼梯间连接不同的空间，并生动地展现了到体育中心锻炼的运动员的进出情况。立面止于建筑底层的一条长凳处，而这样的设计将城市街道具有亲和力的公共环境提升了一个新的维度。后侧的立面由不同色彩的铝板拼接而成，以匹配建筑内部场地已有的颜色。它为体育大厦块状的建筑体量增加了特色，同时也使其形成一个整体，因此协调了内部区域分散的形象。

Groningen Sports Block

Violins street, Groningen, is a fine-grained city street which is now the site of a relatively self-enclosed complex with two sports halls. The potential conflicts between the zoning plan and the functional specification dictated a multistory solution. Alignment of the block of two sports halls proved to have several advantages. It resulted in the block having a relatively modest impact on the street scene, and made it possible to build the volume detached from the adjacent properties thereby making a less massive impression. Two small squares with a tree – on both sides of the building – strengthen the urban embedding of the complex. An arched gate gives access to the rear area where the bicycle shed is located. Schoolchildren can assemble in these spaces before entering the gyms. The sports block thus adapts without contrivances to the fine-woven urban fabric.

东北立面 north-east elevation

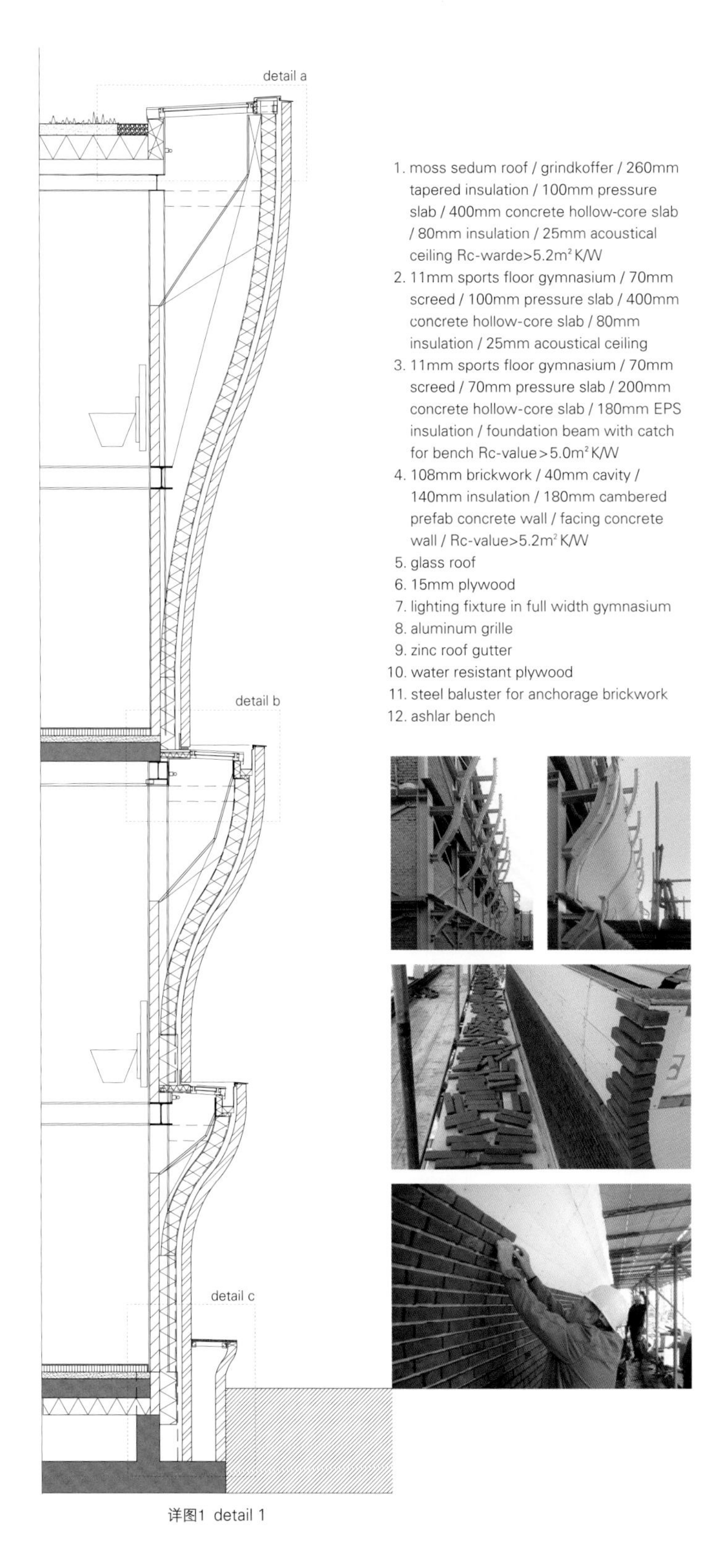

详图1 detail 1

1. moss sedum roof / grindkoffer / 260mm tapered insulation / 100mm pressure slab / 400mm concrete hollow-core slab / 80mm insulation / 25mm acoustical ceiling Rc-warde>5.2m² K/W
2. 11mm sports floor gymnasium / 70mm screed / 100mm pressure slab / 400mm concrete hollow-core slab / 80mm insulation / 25mm acoustical ceiling
3. 11mm sports floor gymnasium / 70mm screed / 70mm pressure slab / 200mm concrete hollow-core slab / 180mm EPS insulation / foundation beam with catch for bench Rc-value>5.0m² K/W
4. 108mm brickwork / 40mm cavity / 140mm insulation / 180mm cambered prefab concrete wall / facing concrete wall / Rc-value>5.2m² K/W
5. glass roof
6. 15mm plywood
7. lighting fixture in full width gymnasium
8. aluminum grille
9. zinc roof gutter
10. water resistant plywood
11. steel baluster for anchorage brickwork
12. ashlar bench

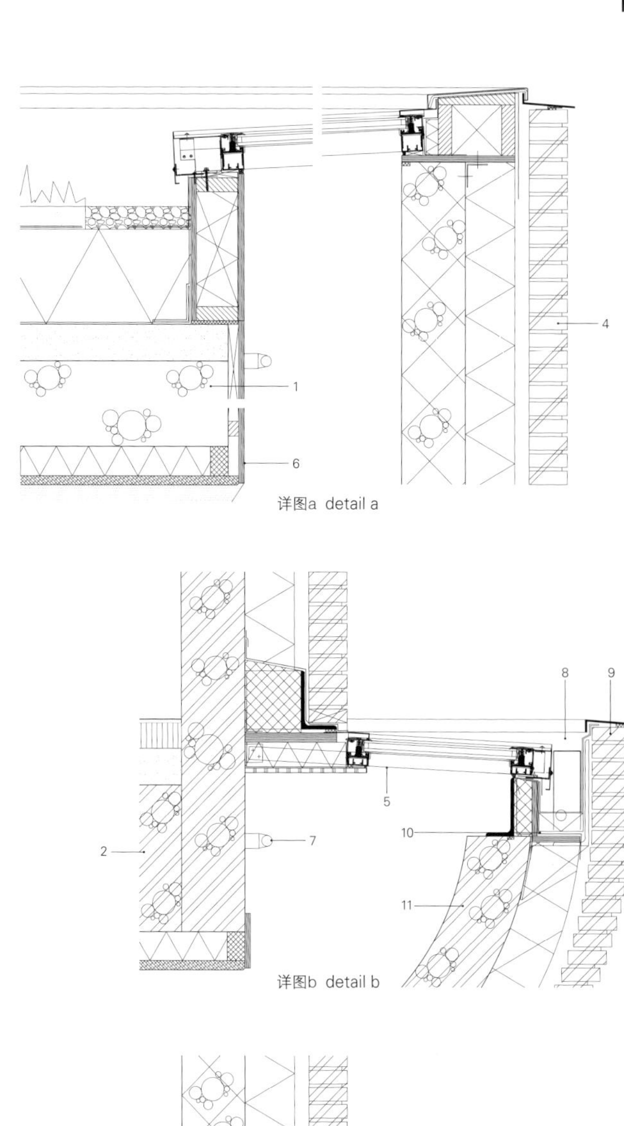

详图a detail a

详图b detail b

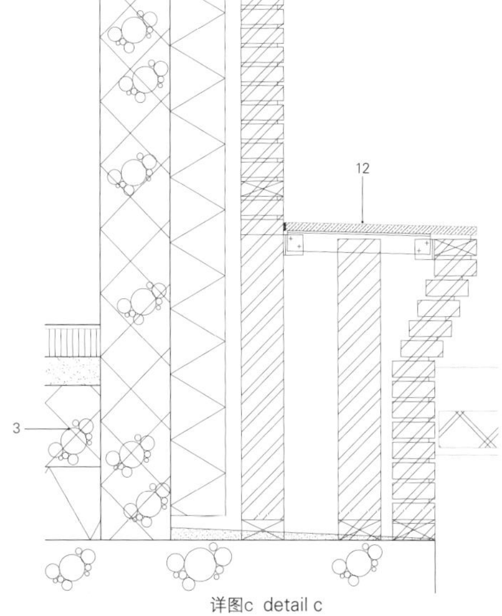

详图c detail c

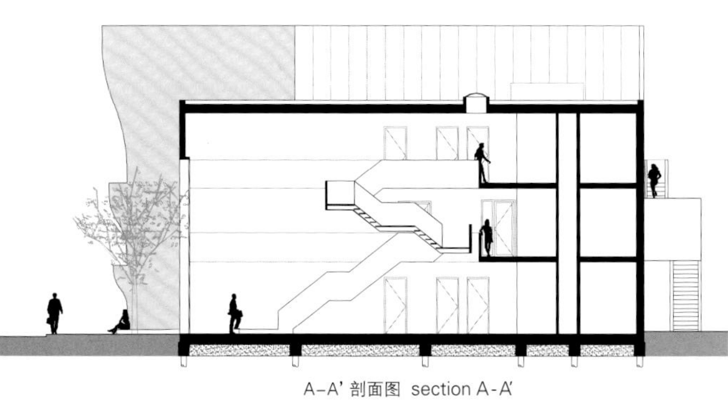

A-A' 剖面图 section A-A'

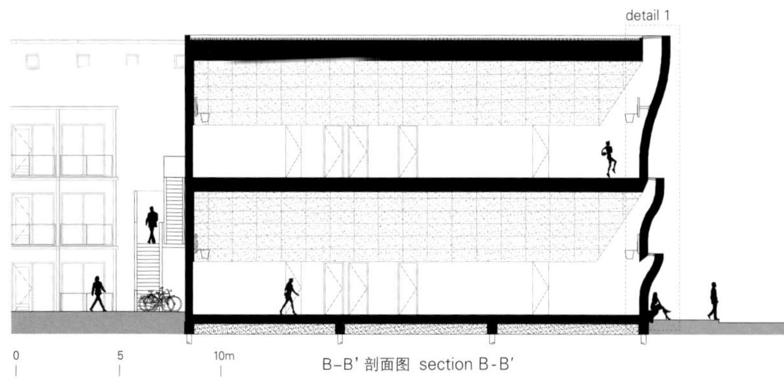

B-B' 剖面图 section B-B'

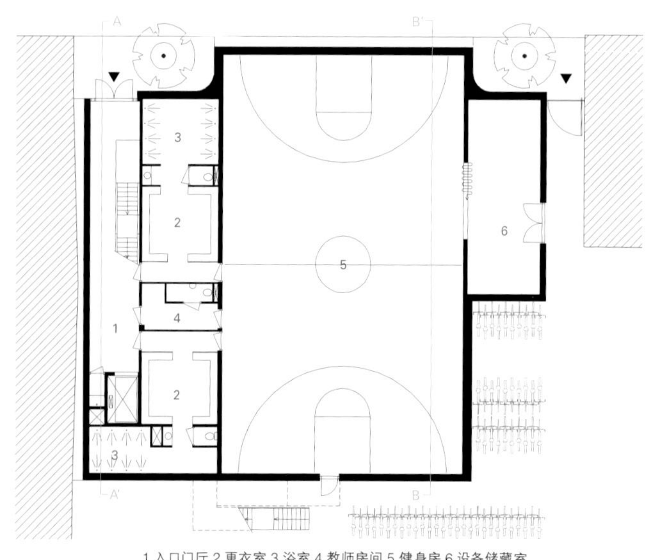

1 入口门厅 2 更衣室 3 浴室 4 教师房间 5 健身房 6 设备储藏室
1. entrance hall 2. dressing room 3. showers
4. teachers room 5. gymnasium 6. storage appliances
一层 ground floor

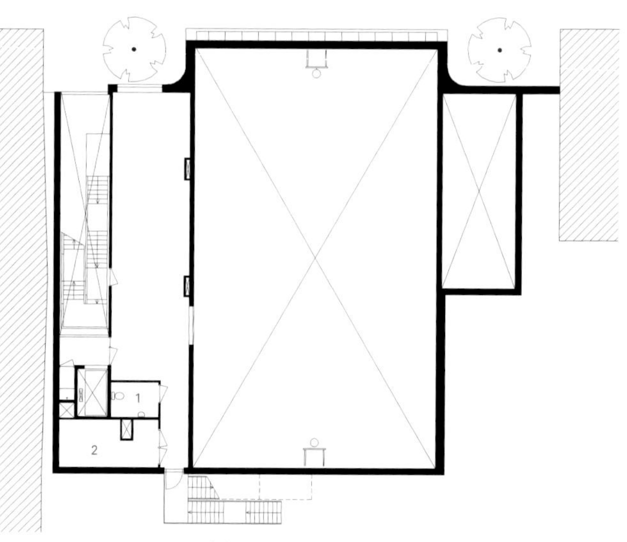

1 残障人士厕所 2 机械设备间
1. accessible toilet 2. technical
夹层 mezzanine

项目名称：Groningen Sports Block / 地点：Groningen, Netherlands
建筑师：Marlies Rohmer
项目团队：Fabian van den Bosch, Ronald Hageman, Thomas van Nus, Begoña Masia Albelda, Cor Martis
承包商：BREED integrated design / 气候顾问：Nieman Raadgevende ingenieurs B.V. / 安装顾问：E&B engineering
客户：Dienst OCSW, Gemeente Groningen
建筑面积：510m² / 总楼面面积：1,139m²
设计时间：2010 / 竣工时间：2014
摄影师：©Daria Scagliola (courtesy of the architect)

A sports hall is in principle an opaque box. Direct daylight is usually regarded as a disadvantage in a sports hall, owing to its potential for dazzling the users and facilitating visual intrusion. On the other hand, operating without daylight is unhealthy because it contradicts the natural biorhythm. The block has therefore been designed with an undulating brickwork facade which admits indirect daylight to each of the sports halls.

The substantial, relatively opaque volume of the sports block was prone to come across as something of an exception to the existing street architecture. It was therefore obviously desirable not to depart too far from the general surroundings as regards the building materials used: an undulating skin of red "Groninger" brick. When the sports halls are used in the evening, the facade glows on using the indirect light lines. A spacious, double height and transparent staircase with a large atrium connects the different spaces and offers a lively image of the coming and going athletes. The facade terminates at ground level in a long bench, which adds a new dimension to the intimate, public-friendly environment of the city street. The rear elevation consists of a patchwork of aluminium panels in several tints that match colours already occurring on the inner site. This adds character to the blocky volume of the sports complex while also articulating it, so harmonizing with the fragmented image of the inner zone.

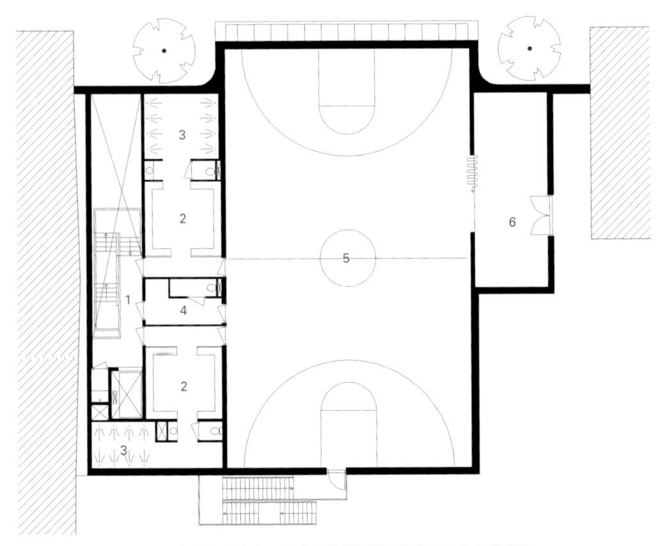

1 入口门厅 2 更衣室 3 浴室 4 教师房间 5 健身房 6 设备储藏室
1. entrance hall 2. dressing room 3. showers
4. teachers room 5. gymnasium 6. storage appliances
二层 first floor

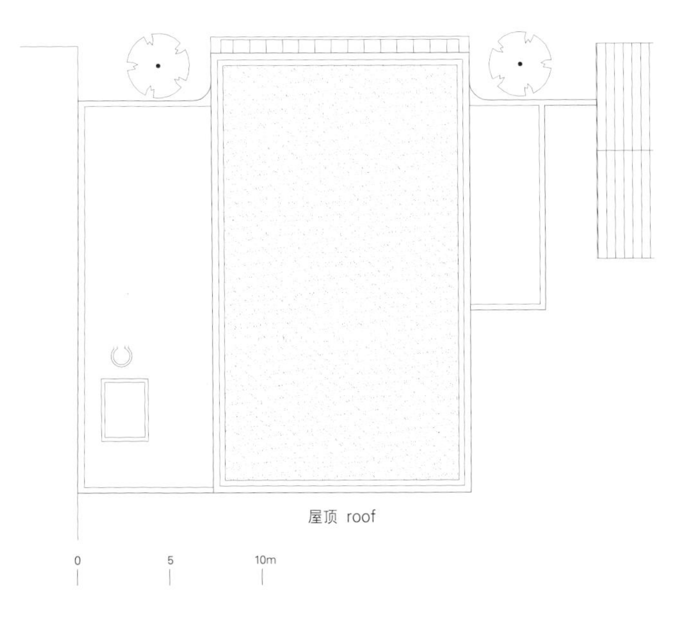

屋顶 roof

在快速的城市化进程中，越南的城市离最初繁茂的热带森林越来越远。例如，在胡志明市，整个城市的绿色植被覆盖面积仅有0.25%。多如牛毛的摩托车导致日常交通拥堵以及空气污染严重。因而，城市地区的新一代人正在失去与自然的联系。

预算仅为15.5万美元的"树之家"，就是为了改变这种状况而设计的一个典型的房屋。项目旨在让绿色空间回归城市，可提供带有大型热带树木的高密度住宅。五个混凝土箱被设计成"盆子"，可在其顶部种植树木，每个房屋都拥有不同的功能。厚厚的土层使这些"盆子"也可以作为雨水池截留和蓄积雨水，因此，如果将来这一概念得以在大量的房屋中采用，将会有助于减少城市的洪灾威胁。

与该场地的不规则形状相配合，五个箱子错落分布打造出一个中央庭院和其间的小花园。这些箱子的大型玻璃门和活动窗提供了面向中央庭院的开阔视野，还可以增加自然采光和通风，同时出于隐私和安全方面的考虑，箱子的另一侧保持相对封闭。常用的功能如餐厅和书房都位于一楼。楼上设有私人卧室和一间浴室，并通过钢制的桥式屋檐连接。庭院和花园被树木间隔遮蔽，也成为生活空间的一部分。住宅模糊的内外边界提供了一个与大自然共处的热带生活方式。

采用当地的天然材料降低了成本和碳排放量。外墙均由采用竹制模具在现场浇筑的混凝土制成，内墙使用当地采购的砖作为终饰。建筑墙体还设有中空层分隔混凝土和砖墙，以减少内部空间的热量散失。

House for Trees

Under rapid urbanization, cities in Vietnam have diverged far away from their origins as rampant tropical forests. In Ho Chi Minh City, as an example, only 0.25% area of the entire city is covered by greenery. Over-abundance of motorbikes causes daily traffic congestion as well as serious air pollution. As a result, new generations in urban areas are losing their connections with nature.

"House for Trees", a prototypical house within a tight budget of 155,000 USD, is an effort to change this situation. The aim of

树之家

Vo Trong Nghia Architects

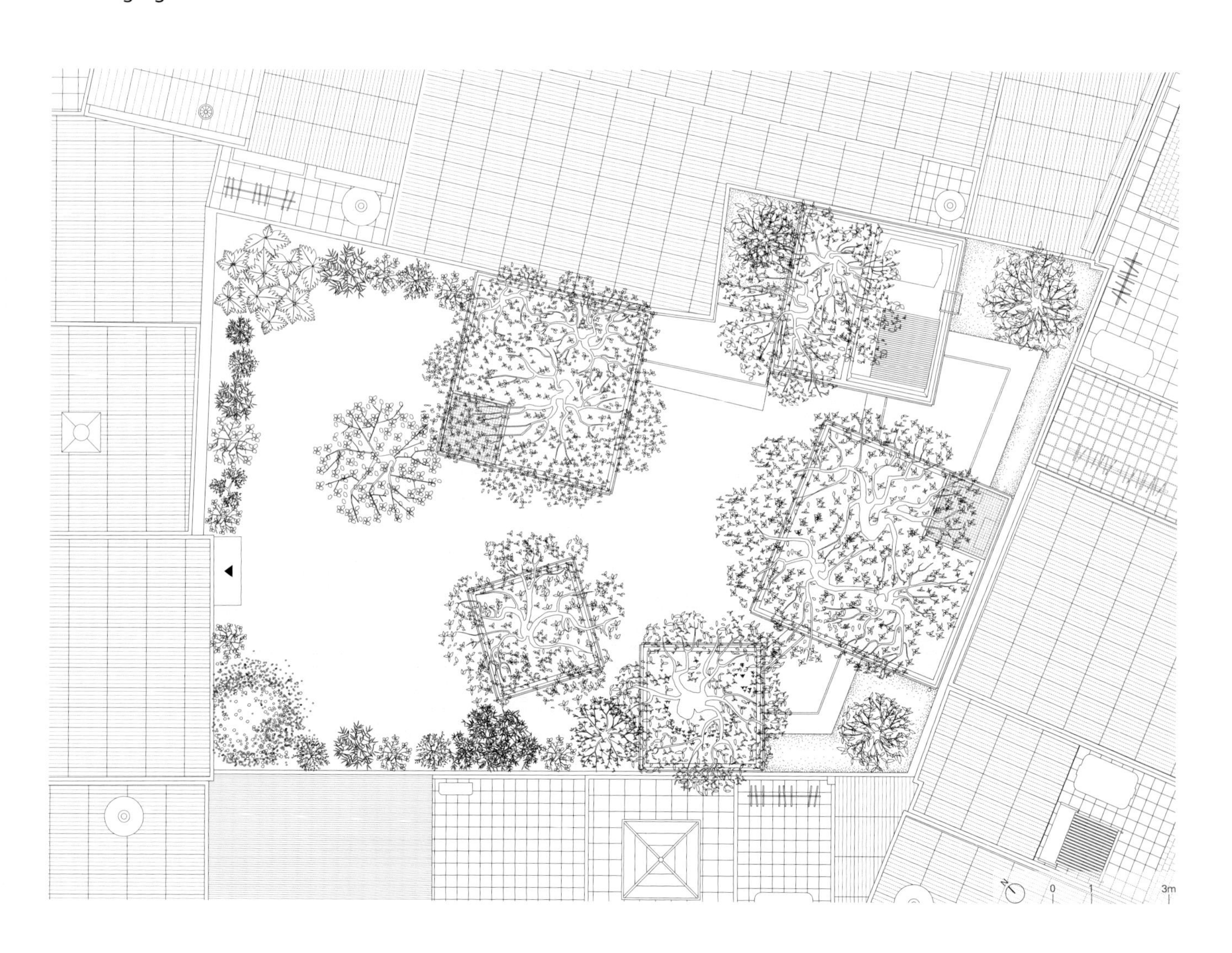

二层 first floor

1 卧室
2 浴室
3 连桥
4 储藏室

1. bedroom
2. bathroom
3. bridge
4. storage

一层 ground floor

1 入口通道
2 储藏室
3 前部庭院
4 中央庭院
5 图书室
6 浴室
7 餐厅
8 厨房
9 祭坛室
10 走廊

1. approach
2. storage
3. front courtyard
4. central courtyard
5. library
6. bathroom
7. dining room
8. kitchen
9. altar room
10. corridor

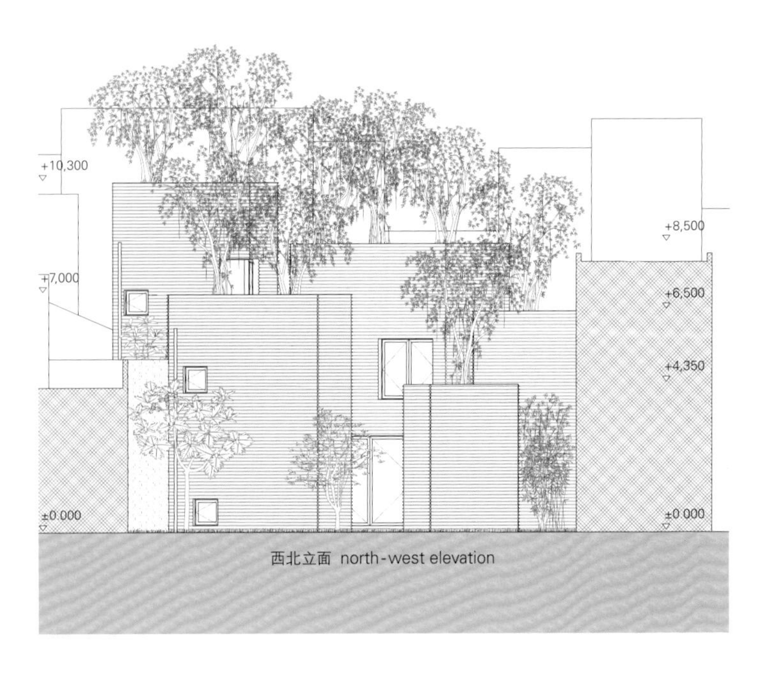

西北立面 north-west elevation

1 卧室 2 图书室 3 祭坛室 1. bedroom 2. library 3. altar room

A–A' 剖面图 section A-A'

project is to return green space into the city, accommodating high-density dwelling with big tropical trees. Five concrete boxes, each houses a different program, are designed as "pots" to plant trees on their tops. With thick soil layer, these "pots" also function as storm-water basins for detention and retention, therefore contribute to reducing the risk of flooding in the city when the idea is multiplied to a large number of houses in the future.

Fitting into the informal shape of the site, five boxes are positioned to create a central courtyard and small gardens in between. The boxes offer open view to this central courtyard with large glass doors and operable windows to enhance natural lighting and ventilation, while remain relatively closed on the other sides for privacy and security. Common functions such as a dining room and a library are located on the ground floor. Upper floors accommodate private bedrooms and a bathroom, which are connected through the bridges-cum-eaves made of steel. The courtyard and gardens, shaded by tree spans, also become a part of living space. Blurring the border between inside and outside, the house offers a tropical lifestyle to coexist with the nature.

Local and natural materials are utilized to reduce cost and carbon footprint. The external walls are made of in-situ concrete with bamboo formwork, while the locally-sourced bricks are exposed on the internal walls as finishing. A hollow layer separates the concrete and brick walls to prevent interior space from heat radiation.

项目名称：House for Trees / 地点：Tanbinh, Ho Chi Minh City, Vietnam
总建筑师：Vo Trong Nghia, Masaaki Iwamoto, Kosuke Nishijima / 建筑师：Nguyen Tat Dat
承包商：Wind and Water House JSC / 用途：Private house
用地面积：474.32m² / 建筑面积：111.66m² / 总楼面面积：226.5m²
设计时间：2012.6~2012.9 / 施工时间：2012.9~2013.6 / 竣工时间：2014.4
摄影师：©Hiroyuki Oki (courtesy of the architect)

该场地是定义建筑及其城市空间的关键。它决定了建筑的城市规模、密度和用途。在该规模和单独的方案中，首先要解决的是分割线模糊的情况，选定的地块带有一个较小的公共立面和一个是其五倍大小的中庭。在这种情况下，建筑被设置为一个独特的墙面交接线的体量；周边环境是经多次改造的住宅区。

在该设计方法中，其挑战在于如何为单一功能构思出一个解决办法：保留具有很高历史和参考价值的地图文献档案馆，并推进对上述的遗产调查和信息披露活动。这些都应该通过一种简单有效的方式来解决。

从使用者的角度来说，我们希望人们以最简单的方式行走于建筑中，获取所需的信息，并且在游历时空间界限越不明显越好。第一和第二个地下室被设计用于地图和历史考察，在这里办公室和与其密切相关的档案共处一室。建筑后方一个英式露台为办公室提供自然光线，而主入口处的两个天窗也提供了自然光。

两百周年纪念项目和大地测量学主管办公室位于中层楼。行政管理部门位于上面两层，设有办公室和对应的档案室。

地面层在该项目中起到了重要的联系作用。它不仅连接不同区域，而且建立了一个兼具功能、活力、多功能的可渗透空间，这一层也因连接了城市和街道空间而得以丰富了视觉延续性，室内空间则因户外花园露台而得以扩充。建筑师将多用途房间设计在该环境的中心，而报告、控制、警示和一般存放区的方位完善了其用途。

人们可以经由宽敞的楼梯间通畅地到达建筑的每一层，楼梯间的灯光采用发出顶棚光的"闪光灯"，在人们经过楼梯间时产生了一种"视觉开放"的效果。

每一楼层的固定点：将电梯、货梯、后楼梯和核心卫生设施设置在建筑的侧面，以优化所有楼层的使用用途和灵活性。

Historical Archive of Geodesia Direction of the Province of Buenos Aires

The plot is the key that defines the buildings and its urban space. It is the one that determines the urban scale, its density and its uses. Inside this scale and its respective scheme, it starts from the assimilation of the parting lines where the chosen land presents a minimum public facade and a courtyard that is five times bigger. In this context, the building is set as a unique body of net lines; pure, surrounded by an environment of residential scale liable to suffer modifications along time.
Within this approach, the challenge would be how to give an answer to a singular program: archives that keep cartographic documentation with high historical and referential value, combined with activities that promote the investigation and disclosure of said patrimony. It should all be resolved in an efficient way and with simplicity.
From the user's point of view, we wish that the building can be walked through to obtain the necessary information in a concise way and traveled through as delimited as possible.
The first and second basements are destined to the department of cartographic and historic investigation, where offices coexist with their corresponding archives which are intimately

布宜诺斯艾利斯大地测量学历史档案馆

SMF Arquitectos

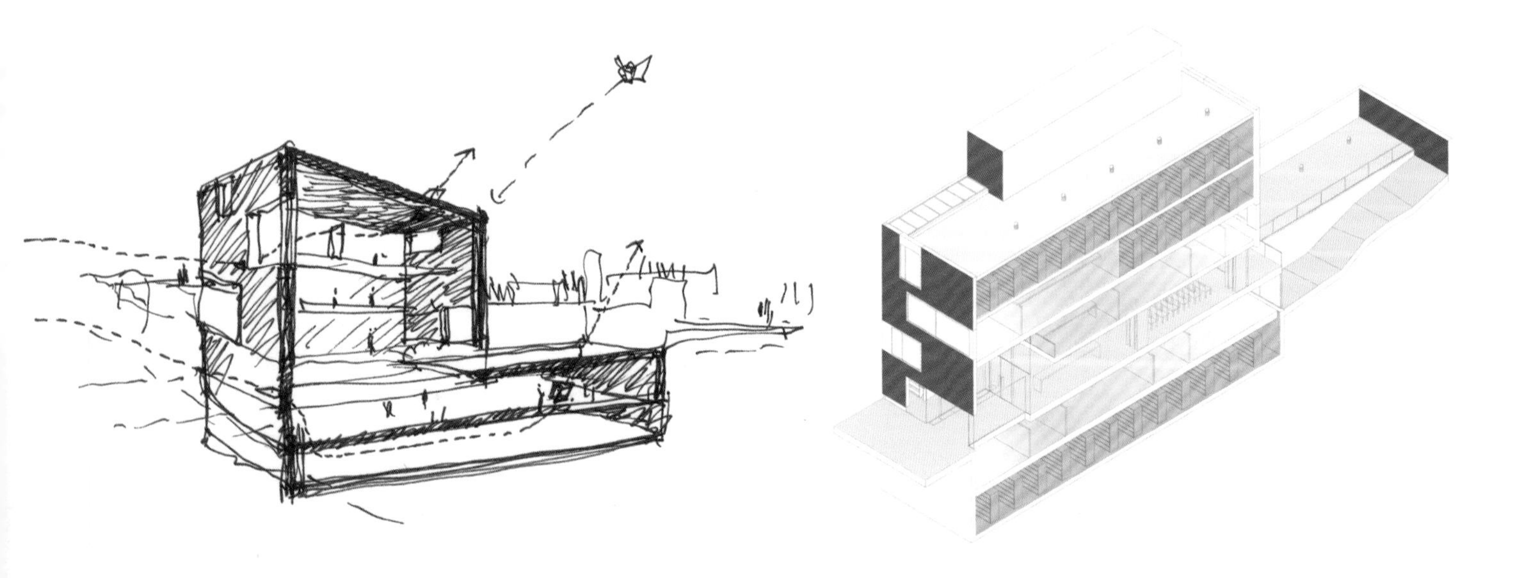

MINISTERIO DE INFRAESTRUCTURA
DE LA
PROVINCIA DE BUENOS AIRES
DIRECCION DE GEODESIA

related. The natural lighting of the offices is obtained from an English patio in the rear part and of two skylights directed to the main entrance.

The bicentenary program and the office of the director of geodesy are located in the mezzanine. The administrative department functions in the two upper plants, with its offices and the corresponding archive.

The ground level is constituted as the great articulator of the program. It does not only serve as binding of the different areas but also builds a permeable space compatible with the function; dynamic, multifunctional, enriched by the continuity of visions that connect the urban, the space of the street, with the interior that is expanded in the outdoors garden terrace. The multipurpose room is positioned inside this environment in a central way and the location of reports, control, caution and general deposit areas complete their use.

A stair of generous dimensions permits the clear access to all the levels where the access of the public is completely necessary and is naturally lit with a "flashlight" that sends zenithal light, producing an effect of "visual opening" while traveling through.

The fixed points in each level: elevators, freight elevator and backstair and sanitary nuclei are placed laterally to optimize the rational use of all the plants and its necessary flexibility.

项目名称：Historical Archive of Geodesia Direction of the Province of Buenos Aires
地点：Calle 61 e/10y11, La Plata City, Argentina
建筑师：SMF Arquitectos
设计团队：Enrique Speroni, Gabriel Martinez, Juan Martin Flores
用途：offices, archive
建筑面积：1,000m²
结构：reinforced concrete
设计时间：2009
施工时间：2010—2015
摄影师：©Albano Garcia (courtesy of the architect)

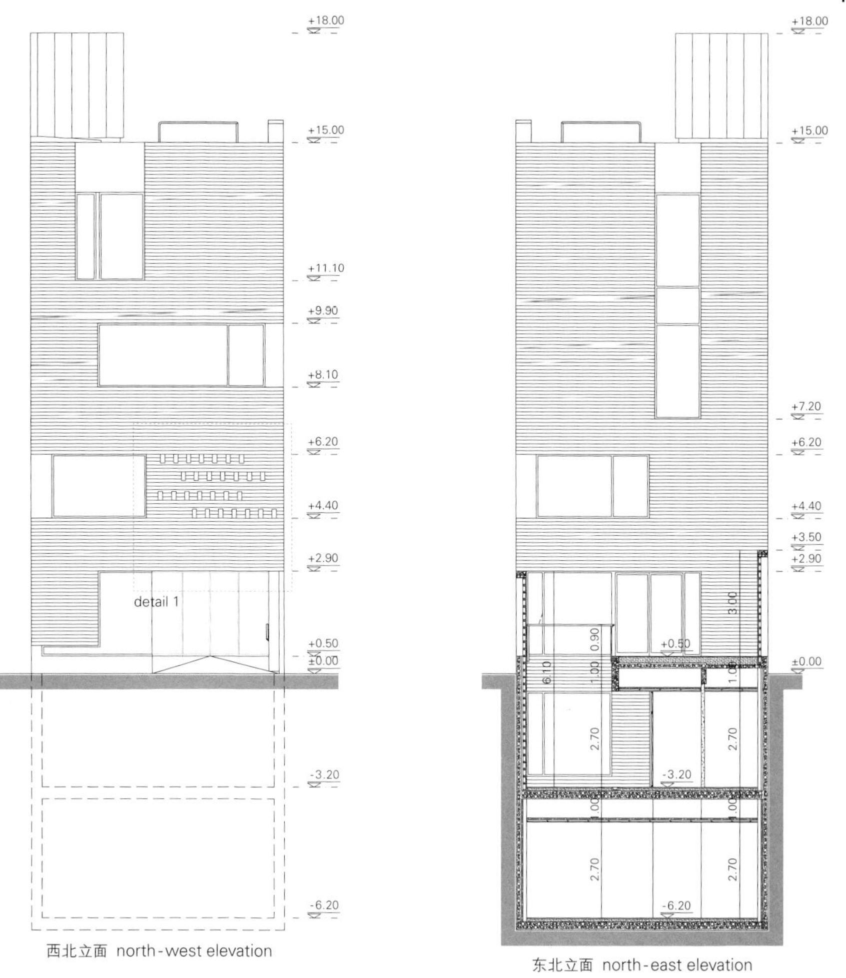

西北立面 north-west elevation

东北立面 north-east elevation

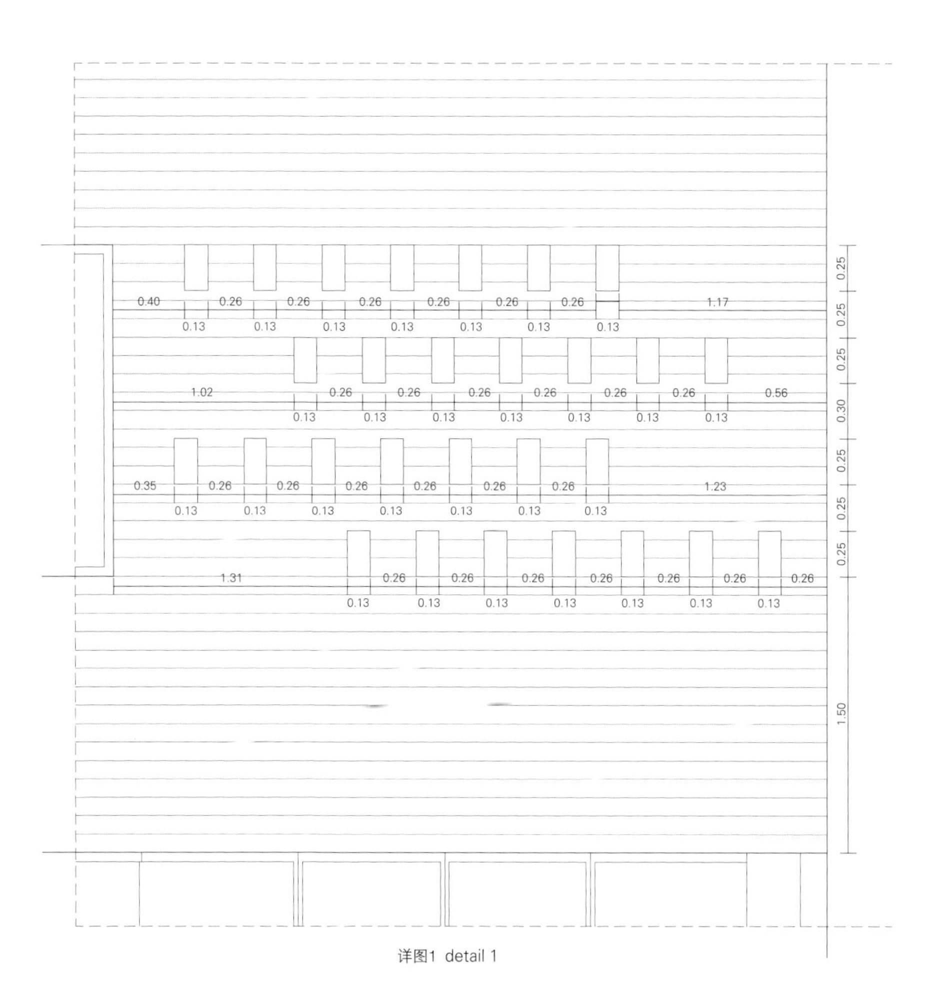

详图1 detail 1

1 发电机房
2 档案室
3 空气压缩机房
4 配电室
5 泵排水室
6 电解室
7 办公室
8 厕所
9 公共电梯
10 储藏室
11 保险柜房
12 充电室
13 电话交换台
14 控制室
15 中庭
16 上方充电室
17 上方入口大厅
18 计算机服务器室
19 屋顶通道
20 发动机房电梯

1. generator room
2. archive
3. compressor room
4. electricity board
5. pumping sewer room
6. tank room
7. office
8. toilet
9. public elevator
10. restore room
11. deposit room
12. income loading
13. telephone exchange
14. monitoring room
15. patio
16. above income loading
17. above entrance hall
18. computer server room
19. roof access
20. engine room lift

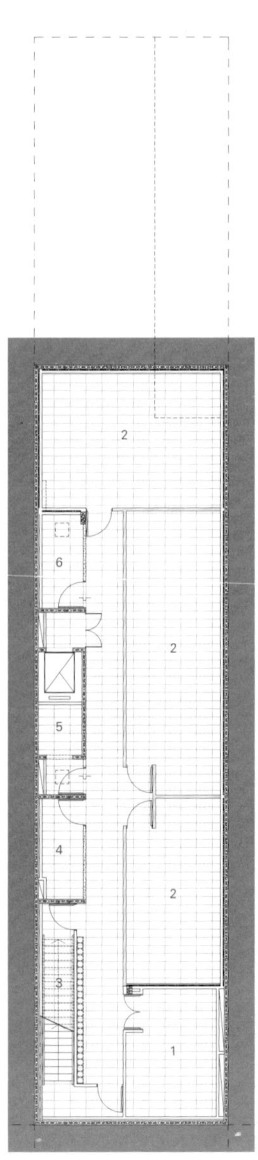

地下二层
second floor below ground

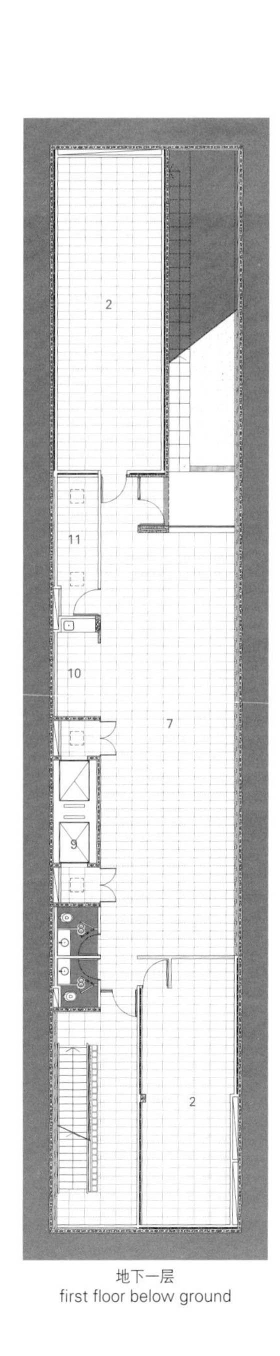

地下一层
first floor below ground

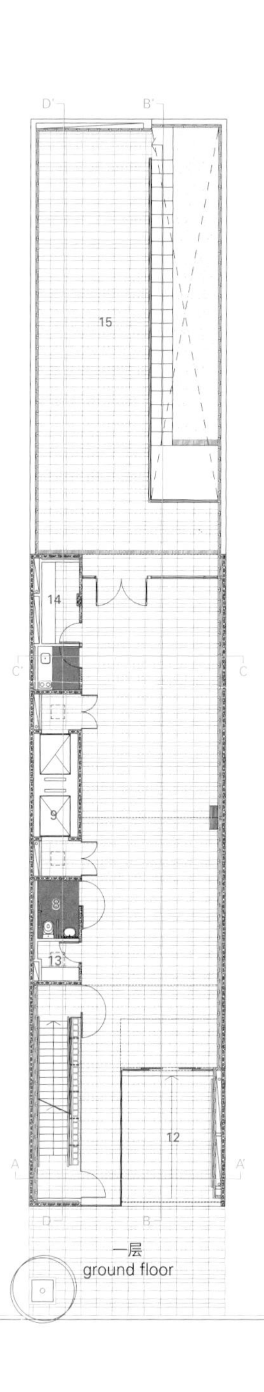

一层
ground floor

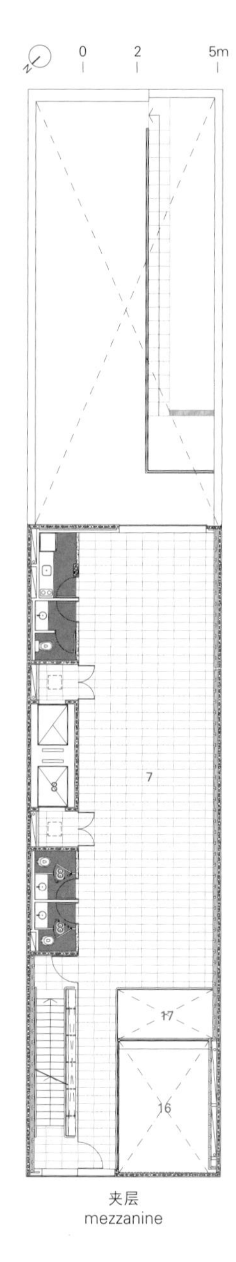

夹层
mezzanine

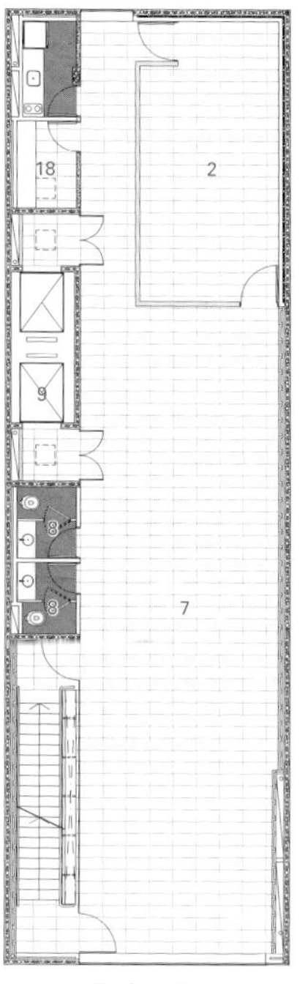

二层 first floor

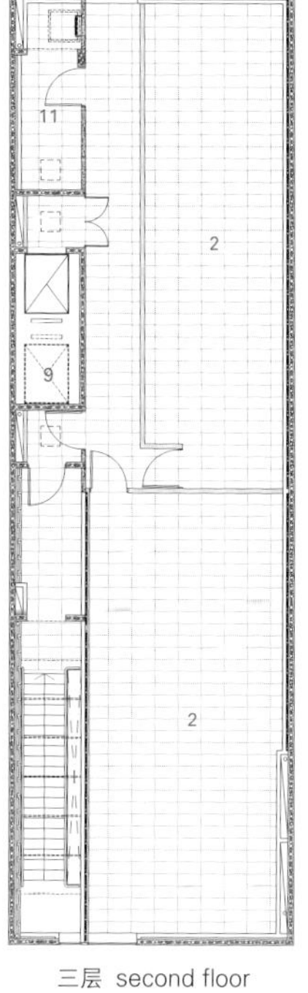

三层 second floor

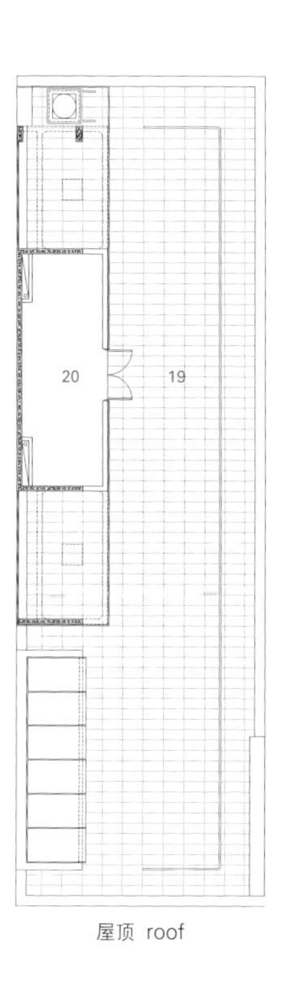

屋顶 roof

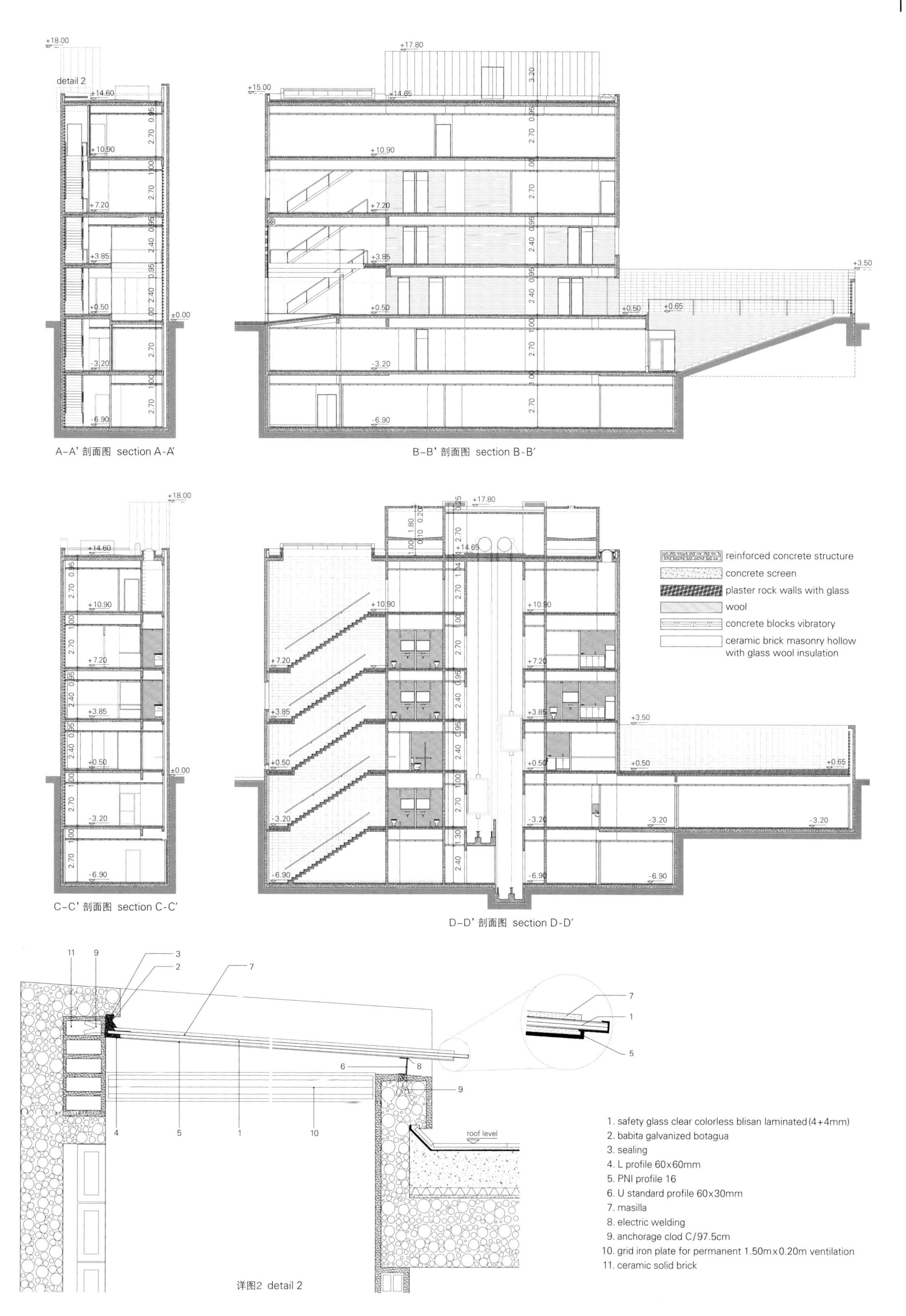

A-A' 剖面图 section A-A'

B-B' 剖面图 section B-B'

C-C' 剖面图 section C-C'

D-D' 剖面图 section D-D'

详图2 detail 2

Luis Anhaia工作室

Zemel + Arquitectos

圣保罗城西的二层住宅小楼被改造成工作室或办公室，每层都有独立的入口。因为该建筑的面积不足150m²，建筑师便决定在开放空间上做文章。

为了获得空间，所有的内墙都被拆除。整个地面都铺上了麻粒岩面砖。巨大的混凝土书架占了整整一面山墙，这一醒目的建筑元素起到了支撑作用。

房子由砖石砌成，因此砖石支撑着山墙、二楼的石板地面和房梁。屋顶也是改造的对象：二楼的天花板被移除，露出桁架。

独立的休息室是必要的配套设施，因此在楼的后面建了一座混凝土塔楼充当休息室。在塔楼的前后各有一块空地，前面的空地铺了方形水泥地砖，后面空地铺了石子，可以在那里吃午餐或聚会。

项目名称：Luis Anhaia Studio
地点：Vila Madalena, São Paulo, Brasil
建筑师：Zemel + Arquitetos
项目建筑师：Paula Zemel, Eduardo Chalabi
合作方：Cristiano Zan
助理：Fernando Milan
施工：Adolfo Droghetti _ Nartex
建筑设备：Antonio Germano Saraiva _ Proan
HVAC：Eng. Miguel Paulo Jardini _ Arconterma
照明：Paula Zemel, Akemi Hizo
供应商：CCL Marcenaria, Camargo e Silva Esquadrias Metálicas LTDA, Esquema Vidros, GR Pisos
用地面积：160m²
建筑面积：83m²
总楼面面积：160m²
设计时间：2012
施工时间：2012—2013
摄影师：©Maíra Acayaba (courtesy of the architect)

Luis Anhaia Studio

A small two-story residential house originally on the west side of the city of São Paulo was bought to be transformed into an atelier or an office, with independent access to each floor. As the grounds have hardly more than 150m², architects made decisions along the line of opening free areas.

To gain space, all of the interior walls were removed and uniform flooring was achieved with a granulite coating. The need of support furnishing was resolved with the highlighting architectonic element of the house: a large structural bookcase made of exposed concrete, which takes up a good part of the side gable and rises from the floor to the roof.

Since the house is made of structural masonry, this element helps sustain the gable itself, the upstairs floor slab and the roofing timbers. The roof was also a target for alterations: the slab above the upstairs floor was removed and the trusses were restored to remain exposed.

Complementing programmed necessities are independent restrooms and break rooms, resolved with the construction of a concrete tower, at the back side of the property. The construction even comes with two outdoors areas, one in front, lined with concrete block paving and another out back, covered with stone gravel and reserved for lunches or informal gatherings.

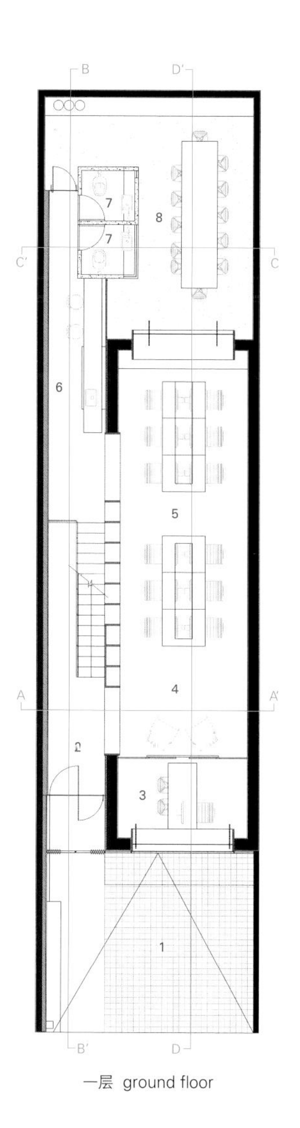

一层 ground floor

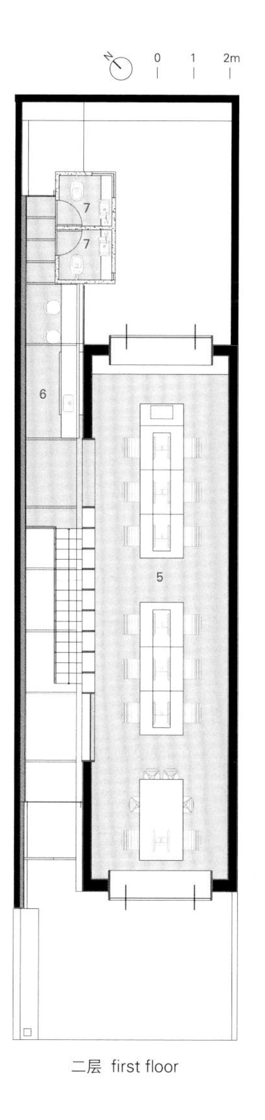

二层 first floor

1 停车场
2 入口大厅
3 行政办公室
4 等候区
5 工作区
6 休息室
7 花园
8 卫生间

1. parking area
2. entrance hall
3. executive office
4. waiting area
5. work area
6. break room
7. restroom
8. garden

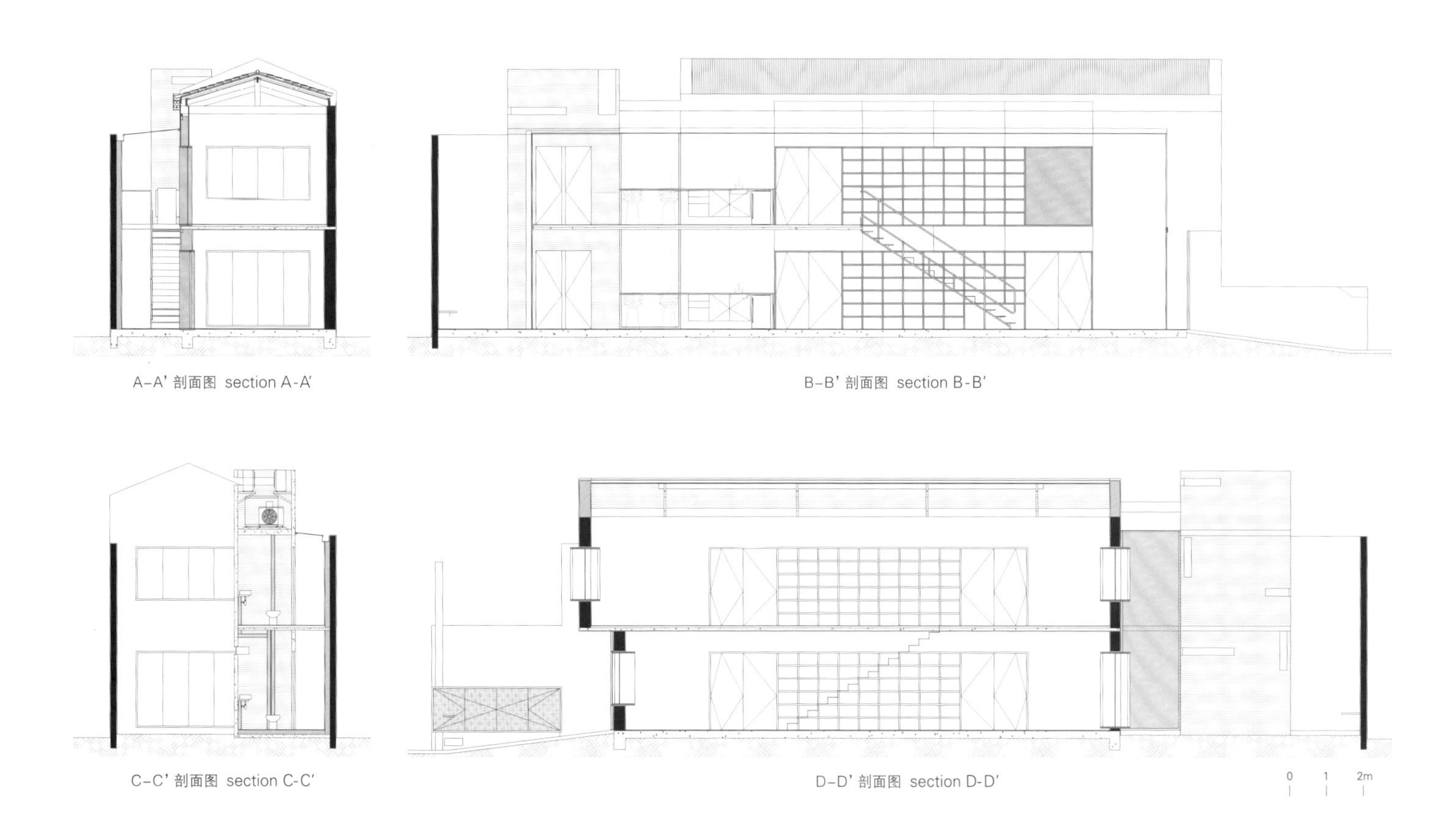

A-A' 剖面图 section A-A'

B-B' 剖面图 section B-B'

C-C' 剖面图 section C-C'

D-D' 剖面图 section D-D'

拉瓦尔德拉玛尔酒店

VVArquitectura

HOTEL
RAVAL DE LA MAR
21

拉瓦尔德拉玛尔酒店位于葡萄牙的比拉塞卡古城中心。古城里的道路纵横交错，建筑大多三层高，被隔墙分成一个个狭长的单元。在这样的城市布局中，排布上的不规则最终强化了当前现有的纹理。

酒店毗邻三条马路，地理位置优越。原址由两栋独立屋和一块空地构成，共有三处房产。酒店由三处房产改扩建而成，形状不太规整。

贝尔赫·皮内达大街在比拉塞卡地位特殊，它连接教堂广场和地中海，途经圣安东尼瞭望塔。街道重要的作用使临街老建筑的正面和后面差异巨大，在后面都是附属建筑和围墙。

在设计酒店的外观时，就要考虑到前面提到的不规整的形状和前后两面的差别。旧有建筑连在一起，它们的外观是有差异的，没有融为一体。因此，我们改变了建筑的外观、垂直切口和上层的护栏，甚至改建了旧围墙的砖石。

经过这些改造，建筑外观焕然一新，从带有历史感的正面（贝尔赫·皮内达大街）过渡到随意灵活的后面（圣约瑟大街）。

酒店的外观低调、普通，与周围建筑协调统一，却与内部设计产生强烈反差。建筑外观隐涩、沉稳；与之相反，内部空间因为采光井的设计而显得宽敞、明亮。

从一层到阁楼，酒店共有36个房间。这些房间围绕着中央采光井呈“万字符”形排列，这样西南方向的阳光能更好地照射进来。上层的外墙向内缩进一点，构成公共或私人室外空间。

酒店的顶层是全属天窗，天窗把酒店内部和外部自然联系起来。天窗的设计与建筑的“万字符”形布局相一致，不仅可以采光，还是整个建筑的组织要素之一。

Raval de la Mar Hotel

The Raval de la Mar Hotel is located in the historical center of Vila-Seca, where the urban lines are irregular and constructions are typically 3 stories high and narrow, limited by dividing walls. In this sort of urban layout, irregularities within alignments have ended up conforming the current existing grain.

The lot has a privileged location since it is delimited by three streets. The plot is the union of three preexist properties on two single-family-houses sit and an empty lot. From the addition of the three we obtain our construction site with its final irregular perimeter.

Furthermore, the Verge de la Pineda Street has a special role in Vila-Seca, it connects the Church's square with the Mediterranean sea, passing through Sant Antoni's guarding tower. This street's importance caused historic constructions to have a facade composition very different from their "back" facade, which were auxiliary constructions and closings.

The irregularity previously mentioned and the difference between the front-and-back facades becomes part of the project when designing the hotel's exterior skin. This skin will break apart and differ where the past constructions met one another. The formalization of this concept are changes on the elevation, vertical incisions in the buildings, changes in the upper handrails or even just the addition of bricks where closing walls had been.

It is through these changes that facades differ and evolve from the front regular facade that reminds of the historic examples (Verge de la Pineda St.), transitioning to a more arbitrary and flexible back facade (Sant Josep St.).

This concept creates an anonymous exterior that blends with its immediate context creating a radical contrast with the in-

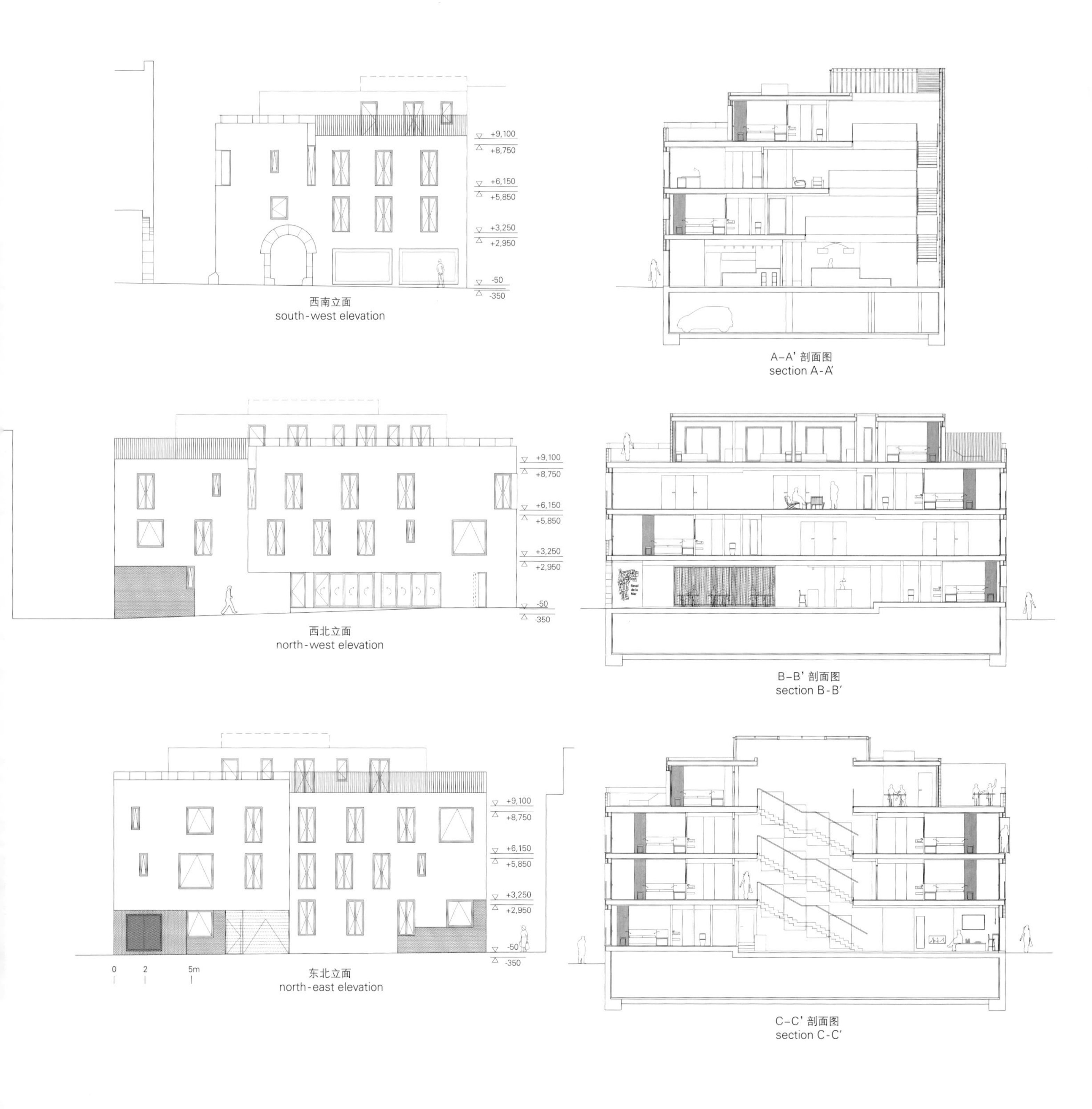

西南立面
south-west elevation

A–A' 剖面图
section A-A'

西北立面
north-west elevation

B–B' 剖面图
section B-B'

东北立面
north-east elevation

C–C' 剖面图
section C-C'

terior design. An opaque and rock-solid exterior that opposes the interior, where spaciousness and brightness are achieved through an interior light well.
The hotel has 36 rooms distributed from the ground level all the way to the attic. They are organized on swastika layout that rotates around the central light well, which tiers to allow a better entrance of the south-west light beams. The upper level's facade steps back to create exterior spaces both for public and private use.
The project reaches its peak with the metallic skylight that connects the interior fortress with the exterior nature. The skylight's structure follows the building's swastika layout and filters the entering light beams, becoming one of the organizing elements of the project.

111

项目名称：Raval de la Mar Hotel
地点：Vila-seca, Tarragona, Spain
建筑师：Alba Vallvè Salas, Pineda Vallvè Salas _ VVArquitectura
技术建筑师：Sergi Balsells
工程师：Josep Maria Delmuns
开发商：Insagra uno S.L.
用地面积：490m² / 总建筑面积：2,250m²
施工时间：2013.2 / 竣工时间：2014.6
摄影师：©Joan Guillamat (courtesy of the architect)

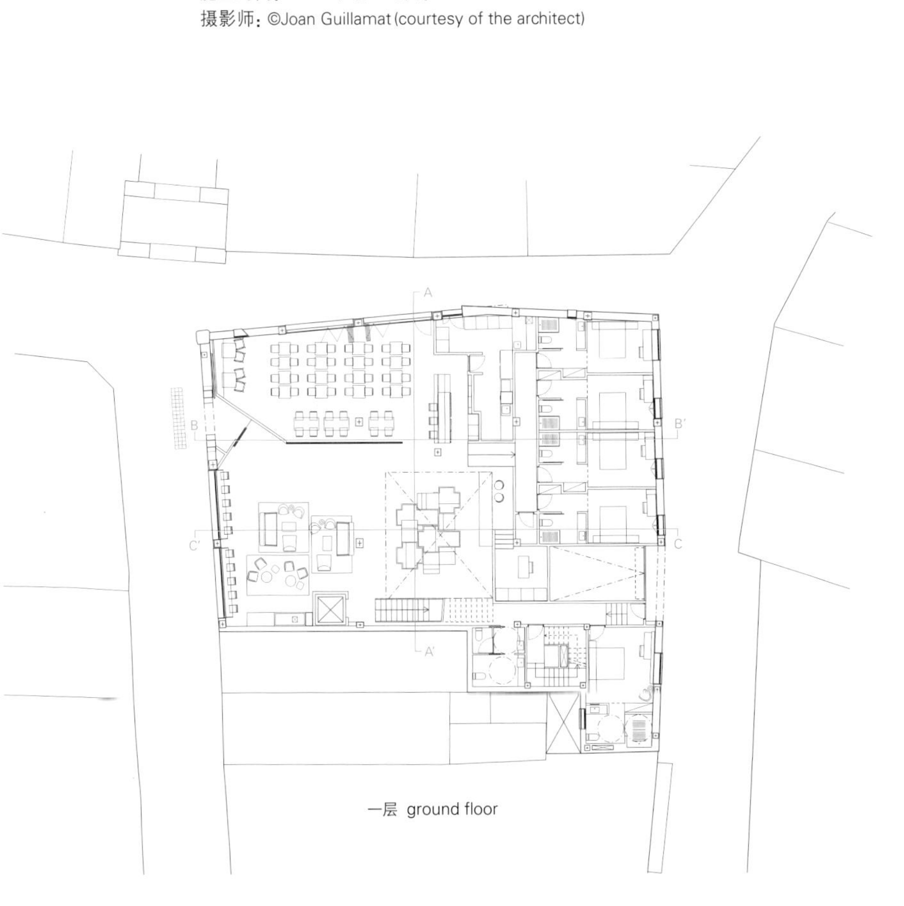

一层 ground floor

屋顶 roof

四层 third floor

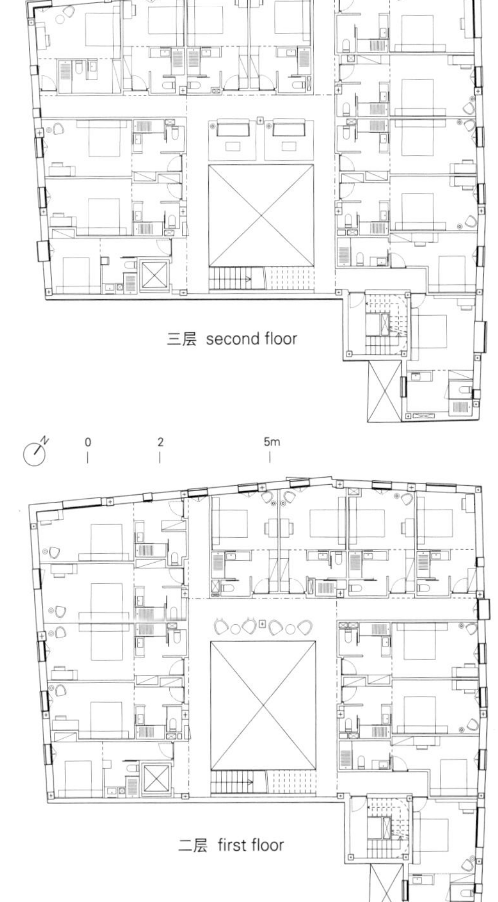

三层 second floor

二层 first floor

211
111

>>100

CO-AP

Was established in 2005 by Will Fung[right] and Charles Markell.
Has realized a diverse range of projects including residential apartments, showrooms, office fit-outs and photographic studios. Will Fung completed his B.Arch with first Class Honors at the University of New South Wales in 1998. After graduation, he has honed his skills at Wiel Arets Architects and Engelen Moore. He has been a regular guest critic and part-time sessional design tutor at the UNSW School of Architecture. Tina Engelen[left] studied interior design at Sydney College of the Arts before becoming Features Editor at Interior Design & Architecture. She co-founded the architectural practice Engelen Moore in 1995 and joined CO-AP in 2007.

>>130

Lens° Ass Architecten

Bart Lens(1959) was awarded a degree in Architecture from the Provincial Higher Institute of Architecture in 1982. In 1995, he set up his own firm, Lens° Ass in Hasselt and opened his second office in Brussels. He became a studio teacher at Sint-Lucas Gent and Brussels recently. He now also lectures at College of Advertising & Design(CAD), Brussels in the Architecture department.

>>164

Vo Trong Nghia Architects

Vo Trong Nghia graduated from Nagoya Institute of Technology with a B.Arch in 2002 and received Master of Civil Engineering from Tokyo University in 2004. In 2006, he established Vo Trong Nghia Co.Ltd.

Aldo Vanini

Practices in the fields of architecture and planning. Had many of his works published in various qualified international magazines. Is a member of regional and local government boards, involved in architectural and planning researches. One of his most important research interests is the conversion of abandoned mining sites in Sardini.

Paula Melâneo

Is an architect based in Lisbon. Graduated from the Lisbon Technical University in 1999 and received a master of science in Multimedia-Hypermedia from the cole Supérieure de Beaux-Arts de Paris in 2003. Besides the architecture practice, she also focuses on her professional activity in the editorial field, writing critics and articles specialized in architecture. Since 2001, she has been part of the editorial board of the Portuguese magazine "arqa–Architecture and Art" and the editorial coordinator for the magazine since 2010. Has been a writer for several international magazines such as FRAME and AMC. Participated in the Architecture and Design Biennale EXD'11 as an editor, part of the Experimentadesign team.

>>120

Rh+ Architecture

Was founded in 2000 by the two associates; Alix Héaume[right] and Adrien Robain[left]. In 2005, they taught at the School of Architecture of Paris-Malaquais as a visiting professor. Awarded by the New 2006 Young Architects Albums Competition in 2006 and also by the Ministry of Culture and Communication allocated to particularly promising European teams of architects for the quality of their projects, the specificity of their career paths and their ability to answer the current architectural and urban issues.

>>186

VVArquitectura

Was created in 2013 by Alba Vallvè[left] and Pineda Vallvè[right] after 7 years' experience including the work at BARQ. One of finalists of III prize at "Alejandro de la Sota" in 2014.
VVArquitectura is characterized by freshness with professional experience. In particular, intergration of history and culture with modern or trendy ideas is extraordinary in their works. They also undertake multidisciplinary projects from urban scale development to interior projects.

>>172

SMF Arquitectos

Enrique Speroni[left], Gabriel Martinez[right] and Juan Martin Flores[middle] graduated from University of La Plata. Their work was focused on preserving the historical heritage but they concentrate more on harmonizing the various local conditions around the neighborhood nowadays. During the first ten years, their work boundary was limited to Argentina. However, they are becoming more visible reference of Latin American architecture in the world.

>>180

Zemel + Arquitectos

was established by Paula Zemel Pompeu de Toledo[left] and Eduardo Chalabi[right] in 2008. They worked together for five years. From 2013 to the present, Paula Zemel has been an architect in charge of Zemel + Arquitetos and Edurado Chalbi has been contributing with studio MK27. Paula Zemel graduated from the FBA-USP(Faculty of Fine Arts - Universidade of São Paulo) in 2002 and contributed with Paulo Mendes da Rocah, Isay Weinfeld and Márcio Kogan, over a period of years. Eduardo Chalabi was graduated from the FAU-USP(Faculty of Architecture and Urban planning-Universidade of São Paulo) in 1999, contributed with Paulo Mendes da Rocha and Bia Lessa.

©JongOh Kim

>>144

Artech Architects

KwanSeok Kim, principal of Artech Architects, studied at Architectural Association, Columbia University Graduate School after B. Arch and M. Arch from Seoul National University, Seoul in 1980. He worked at Ilkun Architects, Zaha Hadid Studio, Christoph Langhof Arkitekten and Kisan Architect etc. before established Artech Architects in 1990. He designed SeJong Government 1-2 Complex with Baum Architects. He received MA Prize for Song Juk Jeong in 2011.

>>156

Architectenbureau Marlies Rohmer

Is an experienced architecture firm, based in Amsterdam, the Netherlands founded in 1986. Has a wide portfolio of commissioned work, ranging from residential to industrial functions and from conversions of existing buildings to complex urban renewal projects. The work is always inspired by research into social and cultural phenomena which are relevant to the project. Marlies Rohmer was born in Rotterdam, the Netherlands in 1957 and graduated from the Architecture/Urban Design at the Delft University of Technology in 1986.

>>80

Arhi-Tura d.o.o

Was established in 2010 in Ljubljana, Slovenia, and now is run by three partners; Bojan Mrežar, Renato Rajnar and Peter Rijavec. Arhi-tura means "a tour in architecture" since Bojan, Renato and Peter traveled extensively during their studies. After graduating from the Faculty of Architecture in Ljubljana in 2004, they worked in different architectural studios in Slovenia, Mexico and the USA before establishing their own architectural studio. One part of their work specializes in architecture made with natural, untreated building materials such as wood, stone, clay, straw bales and hemp.

>>88

Yamazaki Kentaro Design Workshop

Yamazaki Kentaro was born in Chiba, Japan in 1976 and graduated from the University of Kogakuin with M.Arch. in 2002. After graduation, he has worked at Irie Miyake Architects & Engineers. Founded his own architectural design firm Yamazaki Kentaro Design Workshop in 2008 and opened Shanghai branch office in 2011. He believes that the true meaning of architecture is not only a portion of society, but also something that should strive to become a part of our daily life.

图书在版编目(CIP)数据

从教育角度看幼儿园建筑 ：汉英对照 / 韩国C3出版公社编 ；史虹涛等译. — 大连 ：大连理工大学出版社，2016.6
（C3建筑立场系列丛书）
书名原文：C3：Buildings for Kids as Educators
ISBN 978-7-5685-0410-2

Ⅰ. ①从… Ⅱ. ①韩… ②史… Ⅲ. ①幼儿园—建筑设计—汉、英 Ⅳ. ①TU244.1

中国版本图书馆CIP数据核字(2016)第135660号

出版发行：大连理工大学出版社
（地址：大连市软件园路 80 号　邮编：116023）
印　　刷：上海锦良印刷厂
幅面尺寸：225mm×300mm
印　　张：11.75
出版时间：2016 年 6 月第 1 版
印刷时间：2016 年 6 月第 1 次印刷
出 版 人：金英伟
统　　筹：房　磊
责任编辑：张昕焱
封面设计：王志峰
责任校对：张媛媛
书　　号：978-7-5685-0410-2
定　　价：228.00 元

发　行：0411-84708842
传　真：0411-84701466
E-mail：12282980@qq.com
URL：http://www.dutp.cn

C3 建筑立场系列丛书 01：
墙体设计
ISBN：978-7-5611-6353-5
定价：150.00 元

C3 建筑立场系列丛书 02：
新公共空间与私人住宅
ISBN：978-7-5611-6354-2
定价：150.00 元

C3 建筑立场系列丛书 03：
住宅设计
ISBN：978-7-5611-6352-8
定价：150.00 元

C3 建筑立场系列丛书 04：
老年住宅
ISBN：978-7-5611-6569-0
定价：150.00 元

C3 建筑立场系列丛书 05：
小型建筑
ISBN：978-7-5611-6579-9
定价：150.00 元

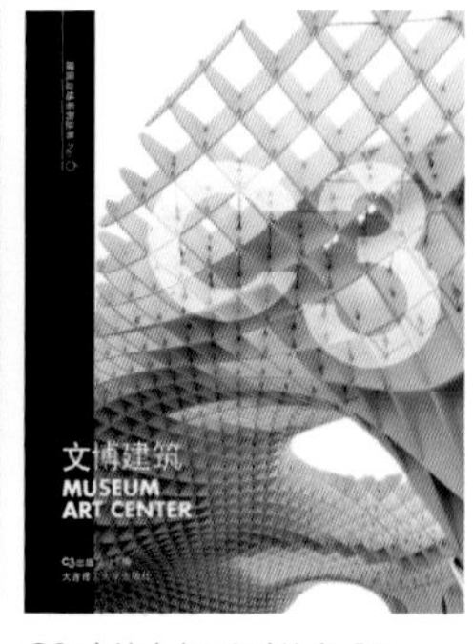

C3 建筑立场系列丛书 06：
文博建筑
ISBN：978-7-5611-6568-3
定价：150.00 元

C3 建筑立场系列丛书 07：
流动的世界：日本住宅空间设计
ISBN：978-7-5611-6621-5
定价：200.00 元

C3 建筑立场系列丛书 08：
创意运动设施
ISBN：978-7-5611-6636-9
定价：180.00 元

C3 建筑立场系列丛书 09：
墙体与外立面
ISBN：978-7-5611-6641-3
定价：180.00 元

C3 建筑立场系列丛书 10：
空间与场所之间
ISBN：978-7-5611-6650-5
定价：180.00 元

C3 建筑立场系列丛书 11：
文化与公共建筑
ISBN：978-7-5611-6746-5
定价：160.00 元

C3 建筑立场系列丛书 12：
城市扩建的四种手法
ISBN：978-7-5611-6776-2
定价：180.00 元

C3 建筑立场系列丛书 13：
复杂性与装饰风格的回归
ISBN：978-7-5611-6828-8
定价：180.00 元

C3 建筑立场系列丛书 14：
企业形象的建筑表达
ISBN：978-7-5611-6829-5
定价：180.00 元

C3 建筑立场系列丛书 15：
图书馆的变迁
ISBN：978-7-5611-6905-6
定价：180.00 元

C3 建筑立场系列丛书 16：
亲地建筑
ISBN：978-7-5611-6924-7
定价：180.00 元

C3 建筑立场系列丛书 17：
旧厂房的空间蜕变
ISBN：978-7-5611-7093-9
定价：180.00 元

C3 建筑立场系列丛书 18：
混凝土语言
ISBN：978-7-5611-7136-3
定价：228.00 元

C3 建筑立场系列丛书 19：
建筑入景
ISBN：978-7-5611-7306-0
定价：228.00 元

C3 建筑立场系列丛书 20：
新医疗建筑
ISBN：978-7-5611-7328-2
定价：228.00 元

C3 建筑立场系列丛书 21：
内在丰富性建筑
ISBN：978-7-5611-7444-9
定价：228.00 元

C3 建筑立场系列丛书 22：
建筑谱系传承
ISBN：978-7-5611-7461-6
定价：228.00 元

C3 建筑立场系列丛书 23：
伴绿而生的建筑
ISBN：978-7-5611-7548-4
定价：228.00 元

C3 建筑立场系列丛书 24：
大地的皱折
ISBN：978-7-5611-7649-8
定价：228.00 元

C3 建筑立场系列丛书 25：
在城市中转换
ISBN：978-7-5611-7737-2
定价：228.00 元

C3 建筑立场系列丛书 26：
锚固与飞翔——挑出的住居
ISBN：978-7-5611-7759-4
定价：228.00 元

C3 建筑立场系列丛书 27：
创造性加建：我的学校，我的城市
ISBN：978-7-5611-7848-5
定价：228.00 元

C3 建筑立场系列丛书 28：
文化设施：设计三法
ISBN：978-7-5611-7893-5
定价：228.00 元

C3 建筑立场系列丛书 29：
终结的建筑
ISBN：978-7-5611-8032-7
定价：228.00 元

C3 建筑立场系列丛书 30：
博物馆的变迁
ISBN：978-7-5611-8226-0
定价：228.00 元

C3 建筑立场系列丛书 31：
微工作 • 微空间
ISBN：978-7-5611-8255-0
定价：228.00 元

C3 建筑立场系列丛书 32：
居住的流变
ISBN：978-7-5611-8328-1
定价：228.00 元

C3 建筑立场系列丛书 33：
本土现代化
ISBN：978-7-5611-8380-9
定价：228.00 元

C3 建筑立场系列丛书 34：
气候与环境
ISBN：978-7-5611-8501-8
定价：228.00 元

C3 建筑立场系列丛书 35：
能源与绿色
ISBN：978-7-5611-8911-5
定价：228.00 元

C3 建筑立场系列丛书 36：
体验与感受：艺术画廊与剧院
ISBN：978-7-5611-8914-6
定价：228.00 元

C3 建筑立场系列丛书 37：
记忆的住居
ISBN：978-7-5611-9027-2
定价：228.00 元

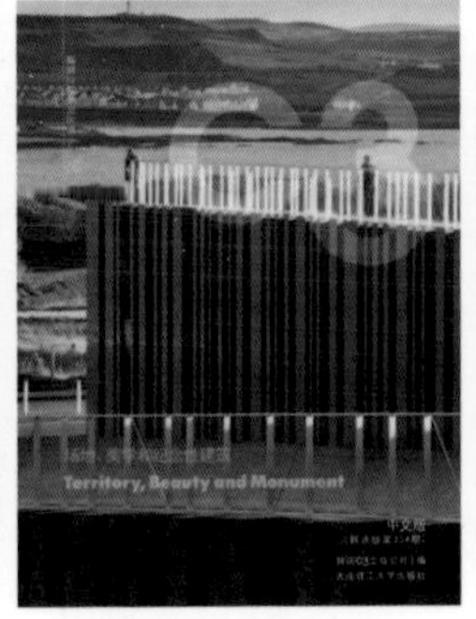

C3 建筑立场系列丛书 38：
场地、美学和纪念性建筑
ISBN：978-7-5611-9095-1
定价：228.00 元

C3 建筑立场系列丛书 39：
殡仪类建筑：在返璞和升华之间
ISBN：978-7-5611-9110-1
定价：228.00 元

C3 建筑立场系列丛书 40：
苏醒的儿童空间
ISBN：978-7-5611-9182-8
定价：228.00 元

C3 建筑立场系列丛书 41：
都市与社区
ISBN：978-7-5611-9365-5
定价：228.00 元

C3 建筑立场系列丛书 42：
木建筑再生
ISBN：978-7-5611-9366-2
定价：228.00 元

C3 建筑立场系列丛书 43：
休闲小筑
ISBN：978-7-5611-9452-2
定价：228.00 元

C3 建筑立场系列丛书 44：
节能与可持续性
ISBN：978-7-5611-9542-0
定价：228.00 元

C3 建筑立场系列丛书 45：
建筑的文化意象
ISBN：978-7-5611-9576-5
定价：228.00 元

C3 建筑立场系列丛书 46：
重塑建筑的地域性
ISBN：978-7-5611-9638-0
定价：228.00 元

C3 建筑立场系列丛书 47：
传统与现代
ISBN：978-7-5611-9723-3
定价：228.00 元

C3 建筑立场系列丛书 48：
博物馆：空间体验
ISBN：978-7-5611-9737-0
定价：228.00 元

C3 建筑立场系列丛书 49：
社区建筑
ISBN：978-7-5611-9793-6
定价：228.00 元

C3 建筑立场系列丛书 50：
林间小筑
ISBN：978-7-5611-9811-7
定价：228.00 元

C3 建筑立场系列丛书 51：
景观与建筑
ISBN：978-7-5611-9884-1
定价：228.00 元

C3 建筑立场系列丛书 52：
地域文脉与大学建筑
ISBN：978-7-5611-9885-8
定价：228.00 元

C3 建筑立场系列丛书 53：
办公室景观
ISBN：978-7-5685-0134-7
定价：228.00 元

C3 建筑立场系列丛书 54：
城市复兴中的生活设施
ISBN：978-7-5685-0340-2
定价：228.00 元

出版社淘宝店

韩国C3杂志中文版已由大连理工大学出版社出版，
欢迎订购！
◆编辑部咨询电话：许老师/0411–84708405
◆发行部订购电话：王老师/0411–84708943

上架建议：建筑设计

ISBN 978-7-5685-0410-2

定价：228.00元